Commission of the European Communities

The Impact of New Technologies on Publishing

Proceedings of the symposium organized by the Commission of the European Communities, Directorate-General for Scientific and Technical Information and Information Management, and held in Luxembourg, November 6-7, 1979

Symposium coordinator:
J. Michel Gibb

Symposium editors:
Marcel Maurice, Edward Phillips, Hans-Ludwig Scherff

Symposium secretary:
Denis Nicolay

K·G·Saur
London·München·NewYork·Paris

Published for the
Commission of the European Communities,
Directorate-General for Scientific and Technical Information
and Information Management, Luxembourg

EUR 6830 DE, EN, FR

(Title of the French Edition: L'impact des technologies nouvelles
sur l'industrie de l'édition;
Title of the German Edition: Die Auswirkungen neuer Technologien
auf das Verlagswesen)

CIP-Kurztitelaufnahme der Deutschen Bibliothek

The impact of new technologies on publishing: proceedings of the symposium / organized by the Comm. of the Europ. Communities, Directorate-General for Scientif. and Techn. Information and Information Management and held in Luxembourg, November 6 – 7, 1979. Comm. of the Europ. Communities. Symposium coordinator: J. Michel Gibb. Symposium ed.: Marcel Maurice . . . — München, New York, London, Paris : Saur, 1980.
(EUR ; 6830 e)
Dt. Ausg. u.d.T.: Die Auswirkungen neuer Technologien auf das Verlagswesen.
ISBN 3-598-10128-7

NE: Maurice, Marcel [Hrsg.]; Europäische Gemeinschaften / Kommission Europäische Gemeinschaften / Generaldirektion Wissenschaftliche und Technische Information und Informationsmanagement

British Library Cataloguing in Publication Data

Impact of New Technologies on Publishing *(Conference), Luxembourg, 1979*
The impact of new technologies on publishing.
1. Publishers and publishing — European Economic Community countries — Congresses
2. Technological innovations — European Economic Community countries — Congresses
I. Maurice, Marcel II. Phillips, Edward
III. Scherff, Hans-Ludwig IV. Commission of the European Communities. *Directorate-General for Scientific and Technical Information and Information Management*
338.4'561'0705094 Z291.7 80-41194

ISBN 0-907150-09-8

K. G. Saur Verlag KG, München
Phototypesetting by film bank, Maria Kerke / Belgium
Printed in the Federal Republic of Germany
by grafik + druck GmbH & Co., München
ISBN 3-598-10128-7

Table of contents

Introduction

by
J.Michel GIBB

Commission of the European Communities,
Luxembourg

Quite a wide range of recent events and developments led to the decision to hold this symposium on "The Impact of New Technologies on Publishing".

It is impossible to list them all but here are a few: the despondency of many learned societies in the face of rising publishing costs, the tragedy of newspaper closures, the anxiety of publishers about the effects of the "photocopying explosion".

Rather than deal piecemeal with each of these troubles, it occurred to us that a global approach would be better. As usual, it is the upsurge of new techniques that is causing the turmoil. As usual, the ultimate problems are human and social in nature.

Any industry involves the cooperation of people with different specialities. The publishing industry in its broadest sense - we might call it the information transfer industry - involves the subtle interaction of an exceptionally large range of people with different skills: authors, editors, printers, booksellers, documentalists, libraries, publishers, etc

Moreover, the various partners in this chain of communication often work in different locations and may not always be sufficiently aware of what the others' role is. New technologies are difficult enough to digest within a closely-knit community. In one so dispersed, they could play havoc.

This is why we felt it our duty to bring together representatives of the different partners involved, on a European scale, in order to help them thrash out the implications, both positive and negative, of technological change, not only as they affect them as members of individual groups, but also as they affect their relationship with members of other groups.

Why do we, at the Commission of the European Communities, feel this to be a duty? There are several reasons.

The basic reason is that the generation, storage, processing and dissemination of information, in all its forms, represents a large proportion of the gross national products of the member states of the European Community (probably something between a third and a half, if one includes education) and that this percentage is increasing as their economies become more sophisticated.Therefore a common market in which proper attention was not paid to information would be incomplete: it is essential that all necessary steps are taken to ensure that information flows without hindrance within the European Community area.

Another reason is the need, apparent in all branches of industry today, to innovate. Innovation is a condition of survival for any enterprise and requires that it not only generate new knowledge itself but also use appropriate information generated elsewhere. The larger the store of information available and the more efficient the methods of securing access to it selectively, the better are the chances of today's entrepreneur. This has necessitated the application of new technology, for instance the creation of communication networks such as Euronet, sponsored by the institutions of the European Communities (see p. 65 for a presentation of Euronet). Now it would be irresponsible on their part to promote new methods and systems, however promising, without also carefully identifying possible adverse consequences of their adoption, in order to be better able to neutralize them.

The collective task of the symposium was thus twofold: first, to understand better how the new technologies, which are available today or are likely to be with us tomorrow, can bring about a more efficient and more effective dissemination of information and, second, to try to perceive the real potential social dangers which their introduction entails and begin to arm against them.

The first session was devoted to a presentation of some of the new technologies and some of their possible applications. Its primary aim was to set the scene for the following sessions, which were devoted exclusively to debate.

In order to save debating time, participants were asked to send in statements before the meeting, as to what they considered to be the major issues and problems and how these could be resolved.

Miss Lydia Gerlach distilled out of the contributions which were received an "issue file" which was the backbone of the discussion session and which is reproduced in Annex I.

In the short time available, it was not possible to do justice to all the issues raised. Perhaps some important problems were even neglected altogether or have emerged in the meantime. Anyone wishing to contribute further to the discussion is cordially invited to send in his views, along the lines suggested in this annex.

Technological developments in the printing industry from now until 1990

M.Yuri GATES

PIRA,
(The research association for the paper and board, printing and packaging industries),
United Kingdom

Abstract

The introduction of new technology is causing rapid changes in the printing industry. The major factors include computers, microprocessors, lasers, digitization of information, screen-based technology including television, and telecommunications. This paper selects those developments and trends likely to be of greatest interest to publishers in the fields of data capture including voice recognition systems, typesetting, graphic reproduction, printing processes, binding and finishing. It also considers non-impact printing processes such as ink jet printing and laser-imaged electrophotographic printing, as well as the effects of the increasing automation of equipment and of the development of alternatives to print, including viewdata and video discs. Implications for publishers are listed.

Introduction

The printing industry, like many others, is going through a period of rapid technological change. The factors which are having a dominant impact are comparatively small in number, but very powerful; they include computers, microprocessors, lasers, digitization of information, screen-based technology including television, and telecommunications. The significant technology changes within the next decade will have their origin in one or more of these areas, rather than in the development of existing technology, and the impact of the changes will be considerable.

The purpose of this paper is to consider what is happening in the typesetting and printing industry now, what is likely to happen in the next decade and what effect this may have on the publishing industry. It is not intended as a comprehensive technology survey, but as a signpost to some of the more significant trends and developments of interest to publishers.

Data capture

Keyboards

Keyboards are, and are likely to remain, the principal means of data capture. The majority of keyboards are built to the standard Qwerty layout, but there are other options; for example, the Linotype 90-key layout used in some newspaper work. Neither of these keyboards is ideal from the ergonomic point of view. The Maltron keyboard is an attempt to challenge the supremacy of Qwerty. It is specially shaped to accommodate the fingers and is designed to speed up the input process by better grouping of the most commonly used characters. However, the Qwerty keyboard is firmly entrenched and will be exceedingly difficult to displace, in spite of its inefficiency

Portable visual display units are available which have a built-in magnetic tape cassette unit. These can be used by reporters, for example, and have full editing facilities. When the story is complete, it can be transmitted down a telephone line using an acoustic coupler.

Various portable chord keyboards have been developed which require two or more keys to be depressed simultaneously to generate a character but these have not yet been used in typesetting applications. They are already, however, being considered for inputting to word processors, which are in turn being linked to typesetting systems.

Optical character recognition

Optical character recognition is a well-established technology which has been extensively used, particularly in the United States, to input editorial text and classified advertisements in newspaper work. The advantage is the rapid transformation of text into machine readable form. Its disadvantage lies in the difficulty of detecting hardware errors and of correcting input errors. Its use in newspapers is declining, but it is likely to continue in use in certain specialist areas, e.g. in directory publishing, where it is a cost effective method of capturing large quantities of data. A further specialist application is the use of multi-font OCR, e.g. for republishing out-of-print works. Multi-font machines are expensive so are likely to be made accessible through bureau services, rather than purchase.

Voice input

Speech recognition systems are of three main types:

Isolated speech recognizer
: The utterance to be recognized, which may be a word or phrase, must be clearly delimited by initial and final periods of silence. Such devices are currently available and are suitable for control and command applications.

Connected word recognizer
: This allows the use of free flowing speech to input a group of words, but the user is constrained to use only words within the predefined vocabulary, and

the group must be clearly delimited by initial and final periods of silence. Devices of this type are becoming available for up to five connected words.

Continuous speech recognizer

This accepts natural, free flowing speech, with no constraint forced on the speaker by the recognition system. Devices exist but are not yet commercially available.

The speech recognition system must be trained to recognize the speakers who use it. Performance is best with single speakers. A typical current system will recognize 50-100 words with a 2% error rate. The largest systems recognize about 1 000 words.

Rapidly decreasing costs of computer storage will assist the development of voice input systems and a voice controlled typewriter with a vocabulary of about 5 000 words may emerge towards the end of the decade. However, it is unlikely that voice input will be sufficiently developed for typesetting applications within ten years, except as a machine control device.

Hot metal

Hot metal systems are becoming obsolete. Their main use is for updating work previously set in hot metal.

Input to photosetting systems

Direct entry photosetters

Direct entry photosetters are comparatively inexpensive to buy and are getting cheaper as the microprocessors they contain get cheaper. They are simple to operate. They can have counting keyboards (both horizontal and vertical counting) which enables hyphenation to be undertaken or mathematics setting to be done. Their major disadvantage is the difficulty and expense of correcting a major error.

The latest devices incorporate visual display units and floppy disc systems; some are programmable to cater for different modes of operation, e.g. inputting, editing, and correcting.

An important development is the interfacing of word processors with direct entry photosetters. Potentially this enables a typist to produce work of graphic arts quality.

Off-line photosetting systems

These consist typically of a number of independent keyboards producing paper tape or floppy disc input for an editing/correcting visual display unit which may or may not be capable of justification and hyphenation. Sometimes a line printer is attached to the VDU as a proofing device. The output from the VDU is a paper tape or floppy disc which goes into the typesetting unit. Compared to the direct entry photosetter, these systems have greater speed and versatility (e.g. more typefaces, more special characters, vertical justification), but they are being superseded by on-line systems.

On-line photosetting systems

Both batch and interactive on-line systems are available. They consist of centralized computing facilities to which (1) data capture devices, including visual display units, OCR devices, paper tape equipment, wire services, etc. and (2) typesetters and line printers can be linked. The command structure for the system is independent of the typesetter, which enables much more complex work to be undertaken. Rigid discs are used to store programs and data. With the interactive systems justification and hyphenation are done continuously, and quite complex page makeup can be done automatically.

Proofing

Line printers of all kinds are used for producing proofs, but do not show the job in true size and format. Matrix printers have been developed which overcome this problem. However, it is often quite feasible and more satisfactory to take a proof from the photosetter itself, e.g. by photocopying.

Proof reading

With text held in machine-readable form, the possibility of automatic proof reading arises. The ideal of completely automatic proof reading has not yet been reached, but various dictionary look-up systems to check the spelling of words have been developed. These tend to require rather a large amount of computer storage space. A different approach is the use of an algorithm to convert any word to a 20-bit code in which form it is stored. Subsequent automatic checking is against this code, which is more efficient in terms of computer storage space than the dictionary look-up systems. Words not found in the computer store are printed out for further consideration.

Photosetters

The original (first generation) photosetters replicate the hot metal process. The second generation machines, including direct entry machines, hold negative film masters of characters on a disc or drum and use optical systems to vary the type size. The third generation machines use digitally stored characters which can be electronically enlarged, reduced, slanted and emboldened. A large number of typefaces and print sizes can be created. Digitization also gives the potential of producing pictures as well as text, and machines which can output both text and pictures are already becoming available. This trend will continue. Fourth generation machines (only one of which is commercially available) use lasers for imaging. Resolution is 1 000 dots/inch and output speeds of about 1 000 lines/min. can be achieved. Better and cheaper laser systems will be developed. Fourth generation machines, like third generation machines, have the advantages associated with digitization - the capability of being linked with computer systems and the potential for producing pictures as well as text.

Full page composition

Electronic full page composition is now available for bookwork. Parameters which specify the page layout, including depth, width, where to place pictures and tables, etc. are entered as well as rules for vertical justification. Page numbers and running heads for left and right hand pages can be inserted automatically, as well as footnotes.

Three systems are available for full page composition of newspaper and magazine pages. The individual pieces of text which are to make up a page are brought together, according to given instructions related to the position and area they are to occupy, and then displayed on a screen. If the result is incorrect, the pieces can be adjusted or moved. Spaces of the required size can be left for illustrations. One system is capable of displaying the picture in position too; it also offers the option of outputting text and pictures on microfilm. These systems are expensive and they have not become widely accepted yet, but they are symptomatic of a new approach to the handling of text and pictures for reproduction.

Graphic reproduction

Text-tone integration

The integration of text and pictures has already been discussed from the typesetting angle. It will now be considered from the viewpoint of graphic reproduction.

The most significant development in graphic reproduction in recent years for monochrome work has been the introduction of scanners using lasers capable of converting continuous tone pictures into digital form. An electronic screen is used to break the picture down into dots; no physical screen is involved. Software is used to convert dot density into halftone values. In simple systems the output from the scanner is dumped onto a magnetic disc which can then be input to a photosetter, thus enabling pictures to be integrated with text. In more sophisticated systems this is done on-line.

At present very fine screening is not possible, because there is a trade-off between quality, which is a function of the number of bits of information data capable of being assigned to each dot, and speed of operation (the more bits of information data, the longer it takes to process them). The current practicable upper limit of screening is about 40 lines/ cm, but the limit will be lifted upwards as costs of computer memory and access fall.

Systems integrating text and pictures are likely to develop rapidly. They offer the potential for increased production speeds and reduced labour costs.

Copy-to-plate systems

The idea of outputting text and pictures directly onto film via a photosetter can be extended still further to outputting directly onto a printing plate. Such systems are currently being developed, including one based on KC film.

Colour scanning

Analogue-type colour scanners have been available for a number of years for making colour separations and are still being developed and improved. Their principal limitation is the necessity for converting analogue colour data into digital form for subsequent page make up. Digital-type colour scanners which are of more recent introduction do not have this limitation. They have greater software capability enabling more complex colour modelling functions to be used, thus permitting more exact colour rendering to be obtained. Digital scanners and their associated page make up systems are costly and are likely to be installed only in the larger printing firms and trade houses. Their advantage is that they reduce the amount of labour-intensive montage or assembly work required, thus reducing costs.

At the other end of the scale, small and comparatively inexpensive analogue scanners are being marketed which are suitable for the smaller printer who wishes to produce colour separations in-house.

Printing processes

Lithography

Lithography is a versatile process used in many types of printing, including bookwork and magazine printing. It has developed rapidly over the last two decades until it is now the dominant printing process. This has been due both to its ability to take advantage of developments in computer-aided typesetting and electronic composition, as well as to its ability to satisfy a market need for reasonably priced colour work.

Sheet fed machines are usually used for short to medium run work while web fed machines are used for longer runs, but there is a trend away from the larger sheet fed machines towards narrow web machines in the larger sizes. The use of these narrow web machines (less than eight-page A4 size) is increasing because they can more easily be used with in-line equipment, such as in-line sheeters and folders. Although the publisher must accept some limitation on page sizes with narrow web machines, he is likely to get better value since the printer can build up a more efficient production system with in-line equipment.

These trends will lead to greater specialization on the part of printers, a decline in the amount of general print work done by printers and an increase in the amount of general print work done in-house.

Gravure

Gravure has two main fields of application - publication work and packaging. It has the ability to print at high speed and give a good result even on relatively low quality paper. It can produce high quality illustrations in black-and-white or colour, which leads to its use in long run magazine and colour supplement work. The major technological problem to be overcome is the high cost of preparing gravure cylinders; if this is solved, gravure could challenge the dominant position of lithography. There are signs that a new system of cylinder-making using laser engraving may be commercially available in two to three years' time, in which case gravure could be viable for print runs down to 250 000 copies. Cylinder preparation seems ideally suitable for EDP application which would lead to simplification of the process, higher speeds and reduced cost.

Letterpress

Letterpress, the original printing process, is in a state of decline. No new letterpress machines are being made and current technological development is directed at extending the life of existing equipment (letterpress equipment is very rugged and long lasting). Flexible photopolymer plates will continue to be used on reel-fed rotary letterpress machines for the production of paperback books where high quality halftone work is not required. Research has shown that pressure distribution in letterpress printing is a fundamental limit preventing it attaining the quality achievable with lithography or gravure.

Binding and finishing

Binding and finishing cover a very wide range of operations including cutting, folding, gathering, sewing, stitching, glueing and covering for the production of books, magazines and brochures. The bindery is the most labour-intensive section of the printing works and for this reason alone is likely to be the subject of considerable development in the next ten years. It is expected that there will be a continuous trend to link the various finishing operations and automate the production lines. There are already two systems in which production is completely automated - paper in at one end, printed-and-bound books out at the other. These systems are good for longish-run products where high quality halftones are not required, but the future trend will be to a more modular approach, allowing book production systems to be assembled to meet individual requirements.

There has been a trend from hardcased sewn books to paperbacks, and from sewn to adhesive binding in hardcased books. There will be continued development of adhesive binding systems and these will continue to replace sewn binding except in high quality books. The introduction of automatically fed and programmed sewing machines will permit cost reductions sufficient to enable high quality books to continue to be produced with sewn bindings.

Non-impact printing processes

Ink jet printing

The general configuration of a typical ink jet printer is shown in Figure 1. A jet is formed by forcing ink under pressure through a small nozzle. The jet is disturbed by vibration of a suitable frequency so that it breaks up into drops which are uniform in size and distance from one another. A charge electrode is placed near the point at which the droplets break from the jet so that each drop will have on it a charge directly proportional to that on the charge electrode. The droplet stream then passes into an electrostatic field, normally maintained by two deflector plates, and the charged droplets are deflected in a direction perpendicular both to the jet axis and to the substrate passing beyond the deflector plate. The distance the droplets are deflected is proportional to their charge, which is in turn proportional to the voltage applied to the charge electrode whilst the drop was forming from the jet. By suitable choice of the parameters governing drop charging and deflection, a drop of ink may be placed anywhere within a narrow band on the substrate passing beneath the jet. Droplets unwanted for printing may be deflected to an ink catcher or gutter and the ink collected in the gutter is then pumped away and recirculated.

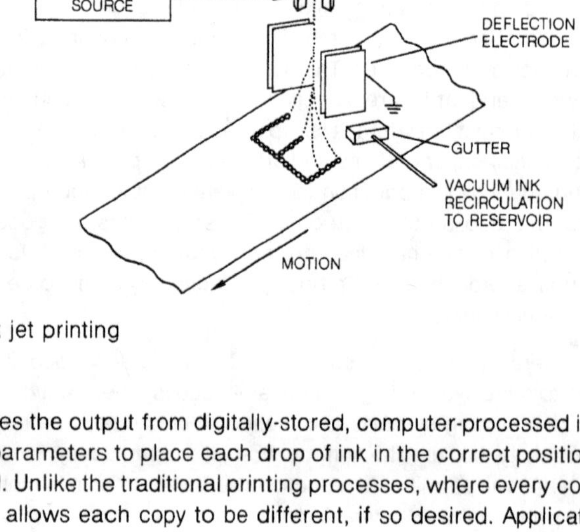

Figure 1. Principle of ink jet printing

Ink jet printing utilizes the output from digitally-stored, computer-processed information, from which the parameters to place each drop of ink in the correct position on the substrate are derived. Unlike the traditional printing processes, where every copy is identical, ink jet printing allows each copy to be different, if so desired. Applications include mailshot letters and personally addressed magazines and newspapers; there are several others, from carpet printing to cheque coding. Potential applications include colour facsimile, colour printout from viewdata terminals, and on-demand publishing.

Some ink jet printers utilize a single jet, some have an array of jets, though most have less than ten. Colour printing is available via multiple jet configurations using three colours. Printing speeds up to 600 m per minute have been obtained. Dot resolutions of up to about 240 dots/inch. have been achieved with commercially available equipment; this is about equivalent to good quality typewriting. The process is capable of further technological improvement and software for half tone work has recently been announced. About 20 000 ink jet printers have been installed throughout the world and it is likely that ink jet printing will become a major printing process.

Electrophotographic and electrostatic printers

Several high speed non-impact printing machines characterized by their ability to accept input from a computer or magnetic tape or disc have become available recently. The speed at which they operate, typically 10 000 - 20 000 lines per minute, is about ten times faster than conventional high speed line printer speed. Working from digitized information, they use laser imaging techniques and are capable of producing print of a quality equivalent to good typewriting and with a varied array of type styles and sizes. Printing is achieved electrophotographically or electrostatically. Further developments will enable line drawings and illustrations to be produced. Possible applications include short run publications, especially those held in machine-readable form for other purposes, e.g. abstracts journals; another possibility is on-demand publication.

Other electrophotographic devices working from digitized data (intelligent copiers) can be programmed to print specialist fonts including Arabic and Kanji. The fonts are software generated.

Colour xerography is already commercially available and a laser-driven colour copier has been developed. However, there is at present a payoff between quality and cost/speed. Technically it is possible to produce high quality colour xerography or electrophotography, but the cost at present is higher than colour photography. The likelihood of high quality colour xerography or electrophotography being cheaper than colour halftone printing for longer runs is small during the next decade.

Automation of equipment

Reference has been made previously to instances of automation of equipment, but the trend is such a major one that it merits special emphasis. Minicomputers and microprocessors are being increasingly used in or with a wide range of equipment, freeing the skilled operator from routine tasks, aiding more consistent results and increasing productivity. Particular examples include electronic composition, the automation of cameras, enlargers and photographic processors, the automation of the setting up and running of the printing press and the automation of bindery operations. Automation of press and bindery operations will have a further beneficial result in that job specification and progress data can be stored and used to pre-programme the individual machines, thus improving work organization. Use of automated equipment by the printer should lead to benefits for the publisher in terms of quality and value.

For the in-plant printer, there is a wide range of equipment available, including photosetters, platemakers, cameras, presses and finishing equipment. Further technical development of this equipment will take place, making it quicker and easier to operate and more versatile in use. It will require considerably less skill to operate than that traditionally associated with printing.

Alternatives to print

The printing industry has survived the introduction of radio, cinema and television, but the main impact of these has been in the field of domestic entertainment rather than information transfer. The rapid development of telecommunications and screen-based technologies such as teletext, viewdata, cable television and on-line information retrieval will constitute a greater threat. Video cassette and video disc technology is just beginning to be marketed and will provide two further alternative media. However, from the publisher's point of view, these new technologies should offer new opportunities and the option of selecting the most appropriate medium for his market.

Conclusions

The graphic arts industries are going through a period of rapid change. Data capture in machine-readable form is becoming easier with the introduction of word processors and direct entry photosetters. On-line photosetting systems are becoming increasingly powerful, offering scope for very complex work to be undertaken, including the integration of text and pictures. Increasing digitization of data and automation of equipment will make printing a more "systems" oriented process and will lead to printers becoming more specialized. The potential for "on-demand" publishing is increasing with the development of digital printing techniques such as ink jet printing and laser-imaged electrophotographic printing and opportunities for facsimile and other telecommunication systems to be used as delivery mechanisms. Alternatives to print, such as viewdata and video discs, are becoming available. Technology is reducing the craft skill required in certain types of printing.

Implications for publishers are:

a. data capture is moving towards the author via the publisher;
b. publishers must increase their awareness of "systems" design incorporating data capture, data-base management and output options;
c. technology will make distributed production and on-demand output possible;
d. there will be an increasing potential for multi-media publishing;
e. there will be increasingly viable options for publishers to produce as well as publish.

Using text processing, computer networking and satellite telecommunication technologies to publish primary scientific and technical information.

John R.U. PAGE

Computer Science Department,
International Institute for Applied Systems Analysis (IIASA),
Laxenburg, Austria

Abstract

The concept discussed in this paper rests on the premise that the conventional system fails to meet the needs for rapid exchange of the results of research and development in certain areas of applied science and technology: conventional publication involves several handlings of the material, is labour-intensive and is therefore slow and expensive. The paper suggests that significant improvements could be made by the use of compatible text-editing systems by authors, referees and publishers, interconnected over computer networks. By creating a master digital record at the outset, many new output options are possible in addition to the conventional journal or report. Examples are new forms of current awareness publications, in the shape of personalized journals and the like, and immediate information retrieval techniques leading to delivery of full texts via the network. To pass full texts through the network will require the use of high-capacity satellite telecommunications, employing small Earth stations. Finally, it is suggested that a programme of studies should be carried out, in association with actual experiments using the European Space Agency's Orbital Test Satellite.

Background: the general concept

The aim of this paper is to present some ideas on the marriage of text processing, computer networks and satellite telecommunications to provide a system for publishing and disseminating scientific and technical information. The basis of the concept is that, where appropriate, texts should be prepared and edited using a standard word processing system. Collaborative publications by several authors not resident in the same locality could be prepared using computer networks (either terrestrial or satellite based) to exchange texts and comments. Refereeing, whether formal or informal, could be dealt with in the same way. The master of the final text would be made available for "publication" in various forms according to need; the bandwidth of the satellite communications system would enable copies of this master to be transmitted in original digital form to different locations for local processing.

Such a concept is admittedly revolutionary, and it must be emphasized at the outset that it is not proposed as a total alternative to traditional forms of scientific and technical publication, i.e. journal articles, monographs, proceedings, and report series and other forms of "grey" literature. The extent to which an all-electronic publishing system could become a complementary or alternative system to present-day methods cannot as yet be forecast: this would require an in-depth economic and operational analysis, and many of the factors cannot yet be quantified.

There are however some pointers which begin to make such a system look attractive for certain classes of scientific and technical information. In 1974, D.B. McCarn of the National Library of Medicine (1) pointed out that the costs of normal publication and distribution were increasing by at least 6% a year, while the cost of data handling and storage of the same material on a computer was falling by perhaps 30% a year. If these trends continued, McCarn concluded, at some point in the relatively near future a cross-over point would be reached at which the routine production and distribution of a formal printed version would not be cost-effective compared with its computer-stored counterpart.

McCarn was referring to the relative cost-effectiveness of consulting the printed Index Medicus as compared with the use of a terminal to retrieve references from the MEDLINE system, but the same kind of relationship would in principle exist in the case of certain types of primary literature. By and large, these trends have continued since 1974, although technological innovation in printing and publishing may have had some effect in holding down the rate of cost increase in that area.

The need for change

The rising costs of conventional publication and distribution appear to be having undesirable effects on the process as a whole, viewed as a means of dissemination of the results of research and development in science and technology. Rising costs have led to increasing subscription fees for journals; library budgets have not increased to the same extent, and in consequence, circulation figures, and therefore revenue, have also tended to fall. Some publishers' organizations claim that the increase in interlibrary loan services, particularly photocopying, has also contributed to this result.

Nevertheless, the volume of material for publication, in one form or another, is apparently still growing. Outside the needs met by monographs and report series, the volume of scientific and technical publication is spread over a very large number of journals and periodicals: one major document supply organization subscribed to over 48 000 journal titles in 1976/77, from which it was able to satisfy over 95% of requests(2).

In spite of the existence of several major serials in many of the main scientific disciplines, a very large number of journals have relatively small circulations. Consequently, it does not seem that the large volume of demand for publication can result in economies of scale.

Moreover, conventional methods of drafting, revising, reviewing, editing and preparing texts for publication are highly intensive in skilled labour and this adds to costs and increases the delay between the point at which a piece of research is completed and the results appearing in print. Even with the most modern methods of handling the final text (computer-aided composition and printing), the time which elapses between completion of the work and its dissemination is rarely less than one year, and in many cases is over two years. Although several interesting new developments have been introduced, such as synoptic journals, with the complete article available as back-up in print, typescript or microfiche, there appears to be no means by which really significant improvements in timeliness can be made within the conventional publication system.

The case for all electronic systems rests on the premise that the number of "handlings" of the material in conventional systems results in high costs and a lack of timeliness which limits their usefulness as a means of information exchange. But how important is timeliness, and what value would the client, in his twin capacity as author and user, place on major improvements in this area?

Hard data, enabling answers to such questions to be given, is not available. It seems clear though that the requirement for timeliness differs between disciplines: in those areas of pure or theoretical science having as yet no immediate industrial or economic applications, timeliness may be considerably less important than in areas of applied science and technology in which the frontier of knowledge is advancing very rapidly and the economic or commercial impact of research and development is more immediate. In certain highly specialized or esoteric areas of pure science, practitioners will be few in number and therefore will probably know each other, so that an effective current-awareness service may be built up on the basis of personal contact and correspondence; a gain in timeliness of formal publication may therefore by irrelevant in the context of information exchange.

Finally, another aspect which must not be forgotten in any discussion of the need for reform of the publication process is that information exchange is only one of the objects of formal scientific publication. Other factors, such as publication being the traditional means of claiming priority for new knowledge, its role as a means by which peer judgement can operate, and (at least in some areas) as a necessary element in career advancement, act as a brake to reassessment of the process as a means of information exchange.

Nevertheless, most users/producers of scientific and technical information would agree that the information exchange role is predominant; the high growth rate of on-line services, and the increasing volume of demand for document copies, are pointers to the growing concern for timely and complete information. Some quantitative details on the relationship of these two factors are given in reference (3).

The technologies: state-of-the-art

In this section we shall very briefly review the current state of development of the three relevant technologies required in realizing an all-electronic publishing concept, together with current moves towards integration to provide a total system.

Text Processing

Text processing constitutes a major growth area in computer applications and the technique has achieved a major breakthrough in commerce and industry during the past few years. Many industrial and commercial organizations find it more cost-effective to buy a mini-computer system with several terminal work stations than to rely on normal typing facilities for the production of reports and memos of all kinds. At the other end of the spectrum, text processing is now replacing other more conventional methods of composing printed matter of all kinds including newspapers.

Aside from the easy manipulation and editing of texts, contributing greatly to cost reduction (no re-keyboarding) and the rapid production of a fully edited master, a feature highly important for scientific and technical information handling is that a digital record is produced automatically during the initial keyboarding operation. This digital record provides a convenient stepping-stone to subsequent retrieval and delivery parts of the dissemination system.

The output of a text-processing terminal or its associated minicomputer/ microprocessor may be interfaced with a telecommunications facility in exactly the same way as any other computer or computer terminal, e.g. they may be connected to a telephone-type circuit (leased or dialled connexion) by modems. Thus, a text drafted in one location may be edited in near real-time in another, either through the public switch telephone network or a computer network. Further, text-processing machines can send and receive texts in this fashion without interrupting the performance of local work at the remote locations. A major manufacturer of text-processing hardware and software has provided quantitative evidence of the explosive growth of interest in communicating systems of this kind (4). Within one year (1978-79), the proportion of systems ordered with communications interfaces rose from less than 10% to over 30%.

While almost every mini-computer manufacturer offers a text-processing system with full telecommunications capability, the systems from one manufacturer are not necessarily compatible with those from another. The need to agree on telecommunications standards has given some impetus to a move towards general compatibility, but there is still a long way to go before texts produced on one manufacturer's system can be edited on another system over a computer network and the revised version displayed on the originating terminal: technically there seems to be no insuperable barrier to such a possibility.

Computer Networks

The expansion and elaboration of computer networking facilities throughout the industrialized world during the last decade is well-known, and needs no underlining here. The major US networks, e.g. Tymnet and Telenet, are now nearly global in coverage: Tymnet extends from the Americas to Europe, the Far East and Japan. After a very slow start, it now seems that thanks to the Euronet initiative, Europe will have its own international public general-purpose data network, interconnecting with national networks within the near future. We are now entering an era in which the emphasis is on network interconnexion, and with the development of the X.75 protocol system, global interconnexion of data networks can only be hindered by administrative and political problems.

Present-day computer networks provide facilties for many types of traffic between computers and input/output devices such as terminals, printers, etc., but most have been designed to meet the special requirements of multiplexing interactive traffic between large numbers of terminals and relatively small numbers of host computers. This is admirably served by the packet-switching technique, which may not however be optimal for the file-transfer type of operation, which requires that a large quantity of data, perhaps of the order of many millions of bits, is transferred rapidly at a steady rate between two fixed locations in the network. Transfer of whole text is, from the computer point of view, more in the nature of a file - transfer operation than interacting with a remote data - base.

Broadly, there are two types of difficulties which could arise in using a packet-switched network for this kind of operation. First, speed of the operation is limited by the maximum speed at which data can be accepted by the circuits entering the main spine of the network (for all practical purposes 9.6 kb/s). Second, a fast steady bit stream between two fixed locations may inhibit the effectiveness of routing procedures, thus degrading the network performance as a whole. The extent to which a given network, such as Euronet, can handle file transfer in addition to interactive traffic is a matter requiring further study and experiment.

Satellite Telecommunications

Two basic properties of satellite communications, as compared with those of terrestrial networks, are particularly relevant to electronic publishing: they are the much greater bandwidth (and therefore bit-rate achievable) as compared with terrestrial data circuits, and the broadcasting, or multi-destination, possibility inherent in transmissions through a satellite link. Geo-stationary communication satellites of increasing sophistication, in the shape of the Intelsat series, possessing these attributes have been in regular operational use for several years, but with little impact in the shape of new or cheaper data services.

This is partly due to common carrier and PTT tariff policies, but is also a function of the technical characteristics of the Intelsat services. In the frequency range used (4 to 6 GHz), and with the on-board power available, very large and costly Earth stations are needed to receive and transmit signals to the satellite. The system is designed for global communications, i.e. for transmission and reception at equal strength over the total area covered by the satellite, a further reason for employing large and powerful Earth stations. Thus, data traffic to and from the satellite must use the terrestrial network to access the single ground station provided within a country or region, and consequently the theoretical benefits of satellite telecommunications are invisible to the data processing user.

However, a new generation of regional communications satellites is now emerging: these are also geo-stationary but concentrate their signals within a particular region using a beaming technique. New parts of the frequency spectrum are employed: the present European prototype communication satellite now in orbit (Orbital Test Satellite - OTS) operates at 11-14 GHz, and this and other design features make it possible to use relatively small and therefore cheap Earth stations with a dish antenna of no more than 3 metres diameter. The cost of such transmit - and - receive Earth stations is at present of the order of 100 000 European Accounting Units, and it may readily be appreciated that with a multiplicity of such stations satellite data communications could become much less dependent on ground-based facilities for local distribution. OTS and an associated Earth station are illustrated in Figures 1 and 2.

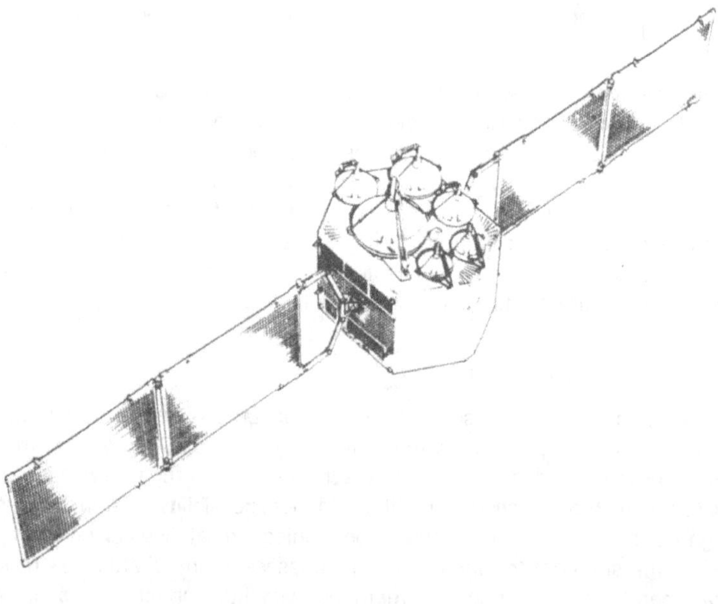

Figure 1: The Orbital Test Satellite (OTS)

Since satellite communications are all digital, there are no special limitations on the rate at which data can be sent and received: rates of a megabit per second may easily be achieved, making massive file-transfer operations much more practical than in the case of terrestrial computer networks. Experimental file transfers using OTS have already been made to test the hardware and software required to interface the computers and Earth stations, with surprisingly low error rates.

The multiple-destination property is also near the point at which practical applications can be tested. The TDMA (Time-Division Multiple-Access) reservation system seems to provide the most economic use of the satellite in this respect; in this, the satellite can accept simultaneous traffic between pairs of Earth stations together with traffic "broadcast" from one Earth station to several others. This possibility is clearly relevant to the requirement for simultaneous text distribution to a number of locations.

Reference (5) discusses communications satellite technology in terms of the various informatics applications now under active study or development.

Figure 2: Typical Earth station for use with the Orbital Test Satellite

Integration of the Technologies

The impact of satellite communications on informatics is receiving a great deal of attention in both Europe and the USA; in both regions, the current emphasis is on the need for integrated satellite and terrestrial networks, each portion undertaking the tasks best suited to it.

In Europe, however, there is not yet total agreement on how satellite systems can best be integrated into the existing networks and services, although there is increasing realization that the advent of the small Earth station opens up entirely new possibilities.

In the USA, the DOMSAT programme, together with the move away from telecommunications monopolies in data transmission, has resulted in proposals by organizations outside the telecommunications field to offer a whole spectrum of new services including electronic mail, video-teleconferencing, high-speed facsimile, text processing and distributed data processing of all kinds. These employ a mix of satellite and terrestrial communications technology and cover electronic data processing applications of all kinds. The proposals of the Satellite Business Systems corporation is a well-known example: in this IBM, in association with COMSAT Laboratories, is offering a large capacity, digital, across-the-board data service using satellites, and Earth stations located on the customer's premises; clients will be large corporations, able to use the capacity offered. The Xerox corporation has recently made a proposal on rather different lines, reviewed in reference (6), which involves a new terrestrial-satellite network to be called XTEN (Xerox Telecommunications Network), composed of roof-top microwave transceivers operating in the 10 GHz band and linked to city network nodes at which frequency conversion to the 4 to 6 GHz satellite band will be made. This mode will constitute the interface with a satellite Earth station. The local microwave radio links will use Frequency Division Multiplexing techniques, and the frequency chosen will permit frequency reuse. It is proposed to offer all types of distributed data processing services on XTEN including teleconferencing backed up by still-frame video and high-speed facsimile. The proposal is currently under consideration by the Federal Communications Commission.

An outline design for an all-electronic publications system

The design of an electronic publishing system can be described in terms of the three main functional elements constituting the publication process in general. These are:

(i) **Origination**, i.e. preparation of material for publication, including any necessary refereeing or approval process.

(ii) **Processing**, i.e. the transformation of the material to a form suitable for widespread dissemination (usually involving text composition or other forms of re-keyboarding).

(iii) **Dissemination**, i.e. the distribution process itself. With present technology, distribution means the printing or reproduction of the text in a journal or as an issue in a report series, and its physical transmission to a point at which it becomes accessible to potential users.

In practice, a fourth process has been found necessary to enable a user to become aware of publications relevant to his particular problem, and therefore the primary literature is abstracted, catalogued, and indexed for inclusion in a secondary publication, designed as a retrieval tool. Almost all such retrieval publications are now fully computerized, and the data bases are available for real-time interactive interrogation using computer networks. The main features and requirements for electronic publishing in each of these areas are summarized in the following paragraphs.

Text Preparation

The main requirement is that compatible word processing systems are used at all stages in the chain, from the production of the first draft by an author up to final amend-

Figure 3: Flowchart for electronic publication

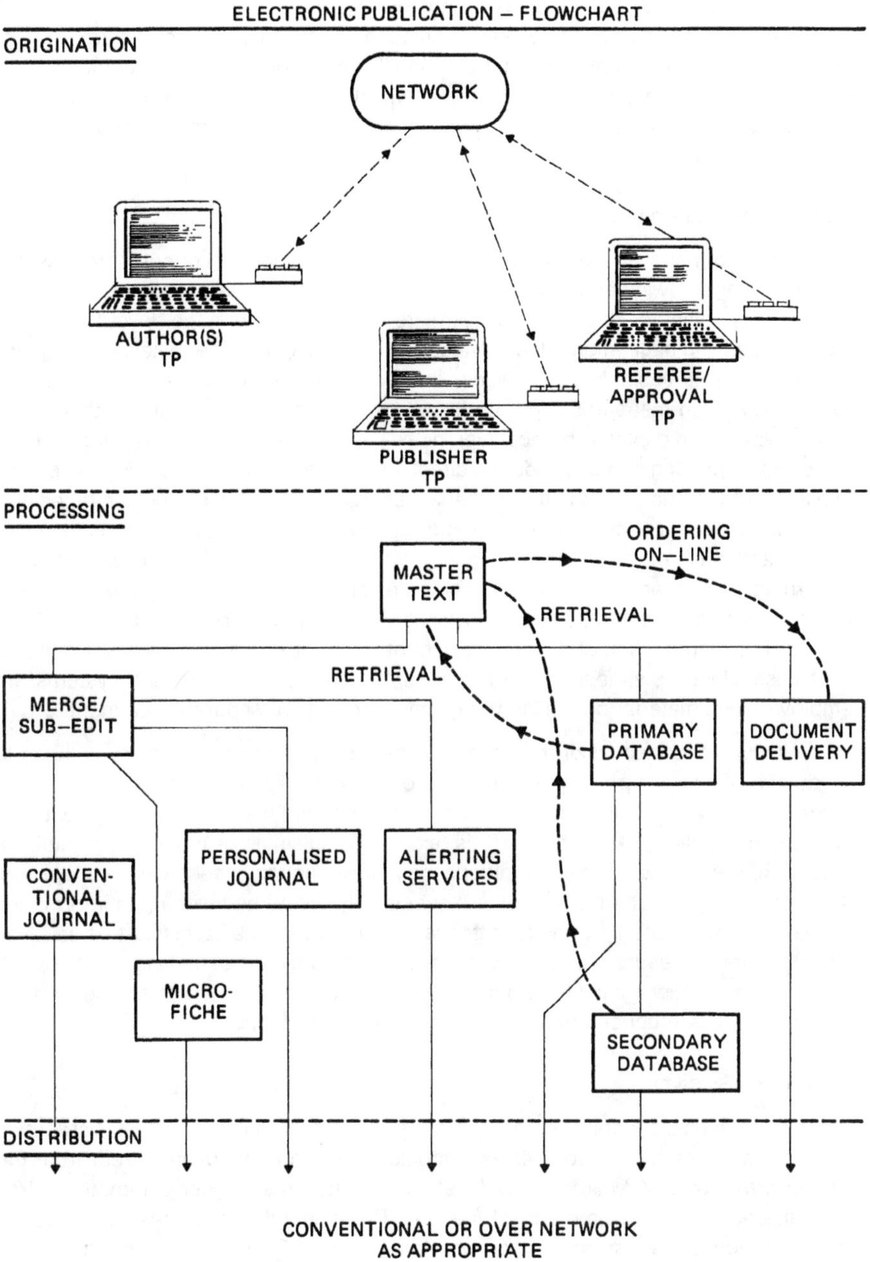

ments as a result of refereeing or approval procedures. This part of the process could work without sophisticated computer telecommunications between the parties involved, relying on the postal transmission of floppy discs or other forms of temporary storage. But, if the costs are reasonable, transmission of the full text using computer networks would be much preferable, bearing in mind that the text is in digital form from the start. Taking into account the rapid growth of present-day terrestrial computer networks, it is not unreasonable to expect that within the next few years authors, editors, referees, and publishers will all have such facilities available, whether using terrestrial or satellite links.

Composition and Reproduction

In the normal case, where reproduction is by some form of printing process, the author's final text will have to be re-keyboarded to provide the necessary master for reproduction, whether this be an offset plate or type. At this stage, layout is usually standardized and pagination added. Although efforts to spread the re-keyboarding load are becoming increasingly common, e.g. the requirement that authors provide camera-ready copy, such measures are not altogether satisfactory. Where the author's final text is available in digital form, however, no re-keyboarding is necessary, since a master for offset printing can be produced directly from the digital record, using the editing facility to add pagination and other material necessary if the text is to be combined with others in one publication. Modern commercial text-processing systems employ clear and attractive type faces which result in good quality masters, far removed from the impermanent appearance of normal computer printout. It may be noted that prestigious journals such as the Harvard Business Review are now produced on a text processor. The reproduction of drawings and photographs using a digital system is at present unsatisfactory, except where highly sophisticated, high resolution systems are employed, and material of this kind may therefore require separate treatment.

Although this discussion has been concerned with using electronic methods to produce a master for printing at a remote location, it may be observed that the existence of a digital, machine-readable form of the text from the outset, together with the capacity of satellite links to deliver this text rapidly to single or multiple destinations, may make the production of an analogue master for printing unnecessary, except possibly for archival or other purposes not specifically concerned with rapid information exchange. If adequate systems for retrieval of information in a particular text are available to potential readers, and if new forms of delivery systems, making use of broadband communications can be introduced as a normal service, then an analogue (print) version could be produced only when and where necessary.

Distribution Systems

Where multiple copies of a printed version are required, normal distribution systems would be employed, with two additional options: microfiche direct from the digital version by a COM process, or local reproduction in a number of remote centres by multi-destination transmission of the text. The most interesting case is, however, that in which transmission and conversion into analogue form is undertaken on demand, either as a result of the particular article or report being retrieved as relevant to a particular user's problem, or as part of a new-style current awareness service.

In the latter concept, users would be able to request new articles matching a particular interest profile at regular intervals (often referred to as the personalized journal concept). In this case too, the output would either be in the form of COM-produced microfiche, or text transmitted to a local printer via a satellite link. In all cases in which full text transmission is required, there would need to be a local delivery system as an interface between the high-speed satellite system and the low-speed local printer, i.e. the production of a low-speed record on tape or floppy disc, for example, which would then be physically transported to the user's location.

Retrieval

If a demand publication system of this type is to be satisfactory, highly efficient retrieval methods must be employed. This would imply the construction of a data-base in parallel with the processing of the text at the publishers. Assuming that an abstract will always form part of the text, it would not be difficult to construct a data-base for free-text-retrieval based on the title and abstract, plus author and corporate source, at the time the texts are being processed for "publication"; this could be done with little or no additional intellectual or manual work as a routine computer task. If indexing were required, additional intellectual work and therefore increased costs and time would be involved. Thus a thesaurus or vocabulary-index data-base could not be made available for searching until the full texts of the articles or reports were available for distribution.

It may be observed that this implies a merging of primary and secondary publication functions, and retrieval data-bases constructed on the basis of what might be described as journal or report series coverage, rather than in terms of material drawn from a number of sources in the same field. It would of course be possible to produce major discipline or mission-oriented data-bases by aggregation of "primary" data-bases, in the same way as is done today, but by speeding up the process using digital records rather than by reworking printed documents. However, such aggregated data-bases could not be produced until after the original documents were available for delivery, in conformity with present practice. Thus, it is probable that, should this form of demand publication be further developed, the optimum design of retrieval systems will require considerable further attention, perhaps in the direction of providing a two-tier facility, the lower for quick retrieval of relevant items from readily identifiable sources, and the second for in-depth retrieval from a large number of sources.

Integration with Other Services

From a telecommunications point of view, the design for an all-electronic publications system sketched here could only become practicable if it were part of a whole new range of high-capacity information and data services offered on a public service basis. The flow chart in Figure 3 and the network layout sketched in Figure 4 should therefore be interpreted in this sense.

Figure 4: Integrated network layout

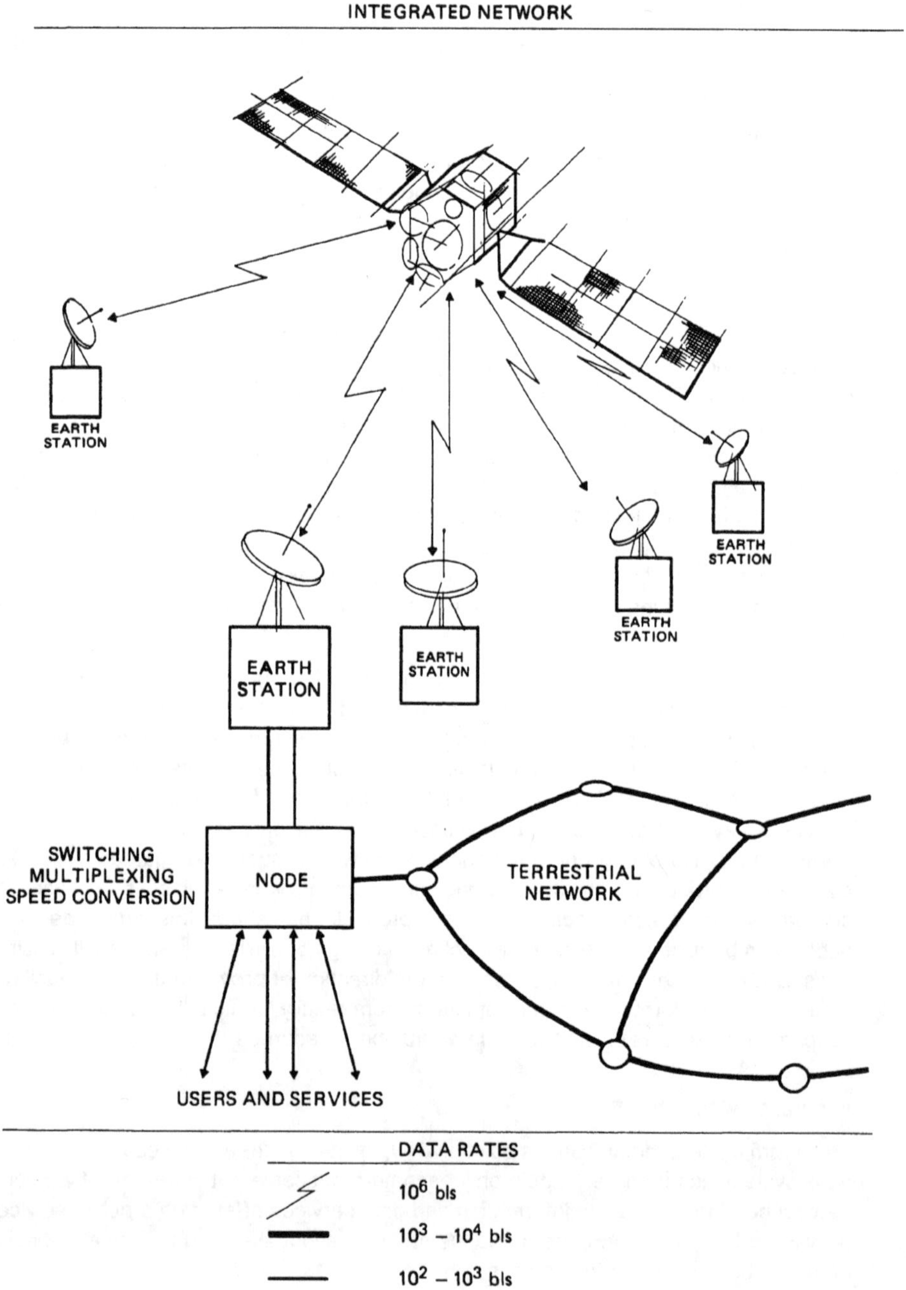

Conclusions

The Requirement

The scheme outlined in this paper is based on the premise that changes are needed in the conventional methods of publishing the results of research and development activity in certain areas of science and technology, in order to improve timeliness and to provide direct access to relevant information. The quantity of material for publication, the labour-intensive nature of conventional systems, the number of "handlings" required between the author's first draft and actual publication, and the time-lag between primary publication and the appearance of tools for retrieving the information, are all factors which increase the total cost of information exchange, and reduce its effectiveness as a total system.

The Possibility of Technical Solutions

The key element in the solution proposed is the creation of a digital machine record at the very beginning of the process as a whole, i.e. the author's draft; this is then electronically manipulated to avoid re-keyboarding and is transported by means of computer networks using, where necessary, the very high-capacity and multiple-destination characteristics of satellite data communications. From a purely technical point of view, text processing, computer networking and satellite telecommunications have all reached a state of development in which the required features are all state-of-the-art, or nearly so. Text-processing systems which can be interfaced with computer networks are already commercially available, although compatibility problems require to be tackled; international computer networks on a public service basis are now at the point at which their universal availability within industrialized regions of the world is beginning to be taken for granted, and the interconnexion of networks is becoming less of a technical than a political and organizational problem. The characteristics of current prototype regional communication satellite systems permit the use of small ground stations and multi-destination transmissions at very high data rates, thus making it possible to transmit whole texts very rapidly.

Operational Considerations

An important feature is the almost total flexibility in the type of outputs which become possible as a result of the digital nature of master texts and their electronic distribution. The conventional options of physically distributed printed versions still remain, but are supplemented by a wide variety of other products. From a systems point of view, the scheme has the advantage that a functional integration of the various parts of the information dissemination and acquisition process becomes possible, each function being tuned to its neighbours in the chain. Under the present system, primary and secondary publication processes are often largely independent, involving costly additional intellectual and clerical effort, with heavy additional resources being required for retrieval based on intellectual and machine reprocessing of material already processed at earlier stages.

Organizational Factors

It seems very unlikely that the technical infrastructure required to operate an all-electronic publications system will be created solely for this application of tele-informatics. However, the new possibilities of integrated ultra-high capacity data networks as vehicles for a whole spectrum of new information and data services of all kinds are now receiving serious consideration, particularly in the USA: a publication facility fits naturally into this more general concept. Indeed, the publication system described could be viewed as a special case of a more general concept for information creation, management and distribution, implicit in the rapid development of tele-informatics.

There are problems in realizing this kind of concept in Europe, but there are grounds for guarded optimism that these may be solved: the collaboration between the Commission of the European Communities and the PTTs in establishing Euronet and between the European Space Agency and CEPT in the OTS programme and its operational successor, the European Communications Satellite system (ECS), together with intense activity in the information and data networking areas at the national level, indicate a rapidly changing climate in which this type of innovation becomes possible.

In the meantime, there are many unknowns, including the whole question of economics. A programme of studies is required to quantify some of the operational and economic factors involved in electronic publishing; this study programme might well be carried out in association with actual experiments using OTS, which is available for applications-oriented experiments in 1980. An experimental programme would enable some of the organizational questions to be resolved without disturbing existing structures. One might, for example, use an existing technical report series in a high-technology area as a guinea pig to assess the overall improvement of timeliness against user reaction to an integrated system involving computer-stored master texts, printed outputs, simultaneous information retrieval and electronic document delivery.

References

(1) McCarn, D.B., "Trends in Information". Proceedings of the 37th ASIS Meeting, Vol. II, October 1974.

(2) "The British Library Lending Division in 1976/77". BLL Review 5 (3) 1977, pp 87-95.

(3) Gillespie, P.D., Katzenberger, P. and J. Page, "Problems of Document Delivery for the EURONET User". Report for the Commission of the European Communities, executed by the Franklin Institute GmbH, Munich, March 1979.

(4) Walshe, W.A., "Electronic Mail Diversifies with Technological Innovations". Word Processing World, April 1979, pp 15-22.

(5) Hanell, S., "The Use of Satellites for Information". ESA Bulletin, 19 August 1979, pp 14-23

(6) "Newsfront".Data Communications, December 1978, pp 15-18.

Information transfer and the significance of new storage media and technologies

Klaus W. OTTEN

Management and R & D Consultant,
Xenia, Ohio,
U.S.A.

Transfer of documented information: technologies and the changing requirements for storage and publication media.

All forms of non-instantaneous communication require documentation: the recording of information on an information carrier which serves as storage medium. In the past, documentation had to be on paper. Paper was considered adequate for all information transfer and storage functions. Increasing demands on transfer speed and growing quantities of documented information reveal inherent limitations of paper, which stem from its low information density and restrict the use of paper for some important publishing and generally for most storage applications.

New media with considerably higher information densities have become available: examples are microforms and magnetic media including tapes, cassettes and discs. They are gaining slowly but steadily in importance because they offer many advantages in the context of computer-supported publishing, documentation and library systems.

This paper develops the rationale for the necessity to apply high density storage media (and thereby also publication media) for most forms of fact, experience and knowledge information transfer.

New information technologies and new storage media have an impact on all forms of distribution of information. Information distribution can be classified into three categories, as illustrated in Figure 1. For the **communication of news,** speed of transfer is one of the most important characteristics. **Information transfer for entertainment** calls for interaction of the recipient with the information. Telecommunication technologies provide almost instantaneous transmission and permit two-way interactive communication. They are already reshaping approaches to the transfer of news and to entertainment. Examples are, besides radio and television, the evolving novel distribution channels created by teletext, videotex and computer networks (forms of electronic publishing).

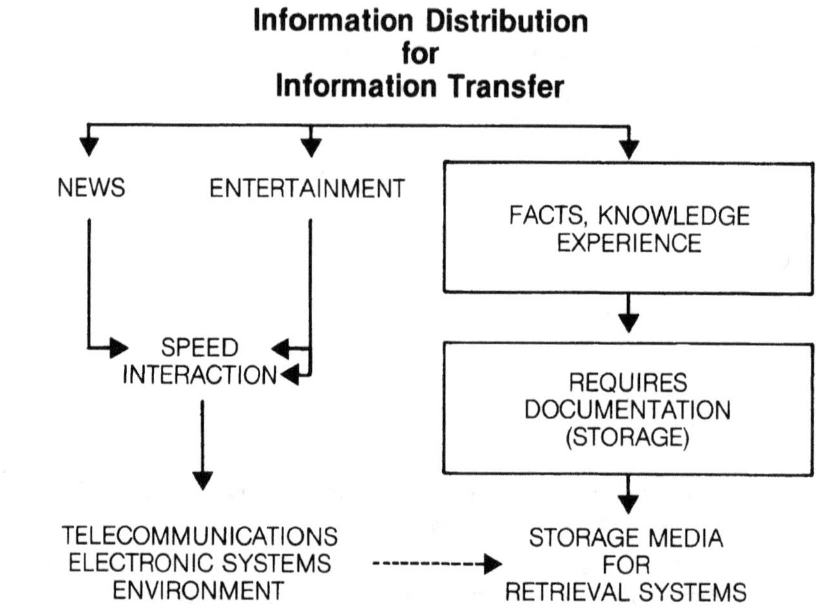

FIGURE 1

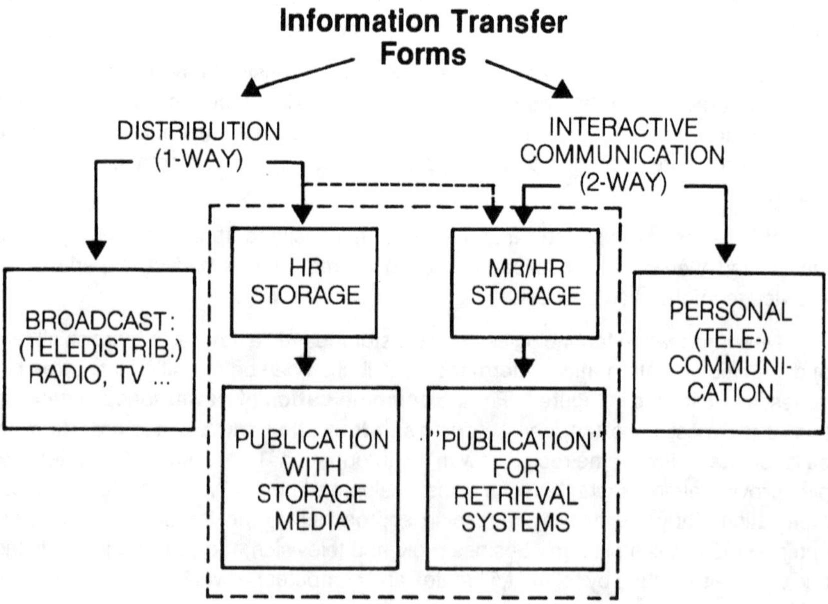

FIGURE 2

Entirely different rules apply to the **transfer of fact, experience and knowledge information**. This information is not transferred for immediate consumption, but rather for storage and retrieval on demand at an unknown later time. It requires documentation in a retrievable form. Therefore the chosen publication method has to be compatible with efficient information management systems and storage.

The significance of storage in the various information transfer modes can be recognized by reference to Figure 2. All non-instantaneous forms of communication depend on storage. These delayed forms of communication can be one-way or two-way with various degrees of interaction. Examples of one-way delayed information transfer are all forms of publishing through, for example, newspapers, magazines, books, phonograph records. Other delayed forms involve interactive information retrieval with the aid of computers, e.g. in bibliographic data-bases.

Forms of instantaneous communication, one-way teledistribution by broadcasting and interactive two-way personal telecommunication do not require storage functions since the information is consumed directly. In contrast, all other forms of communication with delayed consumption necessitate documentation and thereby are based on some form of storage.

In the past all delayed information transfer was exclusively for man as the user: it had to be in human readable (HR) form. Text and pictures on paper satisfied these needs: paper functioned adequately as publication and storage medium. Now that interactive communication modes are being recognized as advantageous, with new information technologies becoming attractive for documented information, the requirements for storage and publication media are changing. Interactive communication involves fast manipulation of documents for retrieval: this is only possible in the context of automated retrieval systems. To permit a dialogue, an electronic and computer supported system is used. It requires indexes, data storage and so on in machine readable (MR) form. Machine readable records can be stored and published on magnetic and some optical media (optical discs, special microforms). Common to all media for MR information is a very high information density in comparison to that of paper.

Automation of document handling also requires high information density, in order to achieve cost-effective mechanization. Paper is consequently unsuited for publications aimed at use in retrieval systems. A combination of storage media for HR information (microforms) and for MR data (on computer storage media) becomes necessary.

Transfer of documented information: changing storage media and technologies

Storage functions are fundamental to the transfer of fact, experience and knowledge information. In the following discussion no further consideration is given to forms of instantaneous communication: transfer of news and entertainment information, for which storage media and functions play only supporting roles.

Storage functions are involved in various phases of the transfer of documented information. They appear to be concentrated in four areas, as shown in Figure 3. For

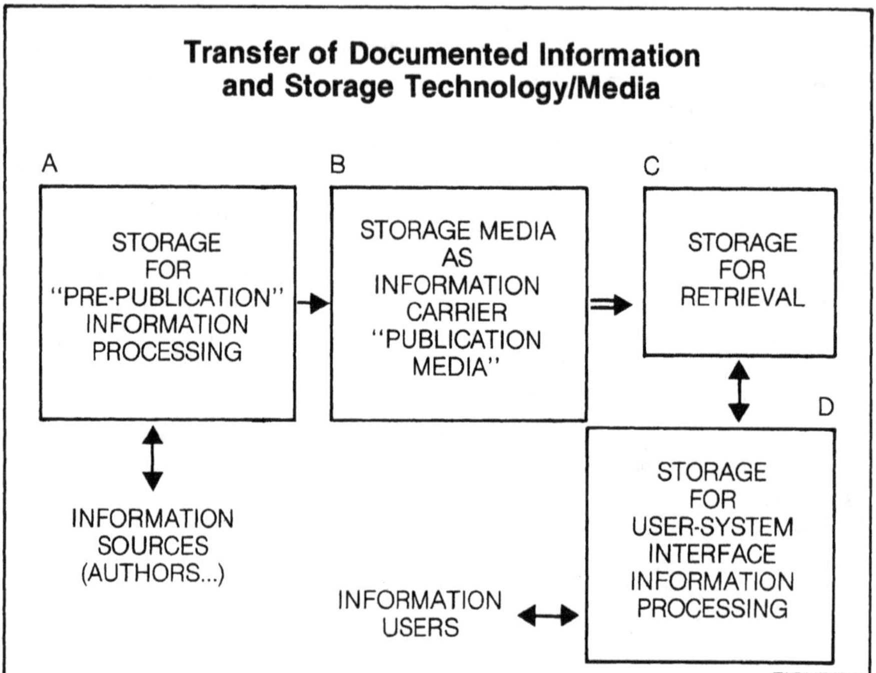

FIGURE 3

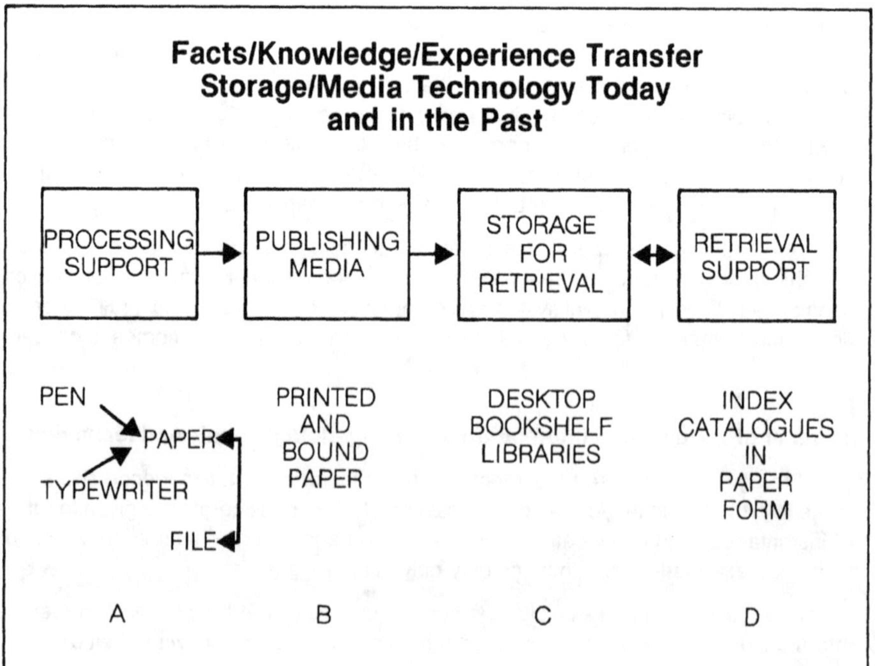

FIGURE 4

each of these areas the functional requirements for storage are different.

For the first area, **pre-publication information processing** (labelled A), a scratchpad-like storage is required. Information should be easily recordable, changeable and erasable. Traditionally pen and typewriter have been the recording instruments, paper the storage medium and files the storage devices.

In the second area, **media act as information carriers** (labelled B). From the standpoint of publishing the media ought to be easily replicable and transportable. Paper served and will continue to serve this function well: it appeals to the end user, because it is easy to handle and it permits convenient access to printed information. But it is now being challenged in certain application areas by high density media, in particular by microforms for reasons of overall cost effectiveness (if costs include filing, storage, file update and retrieval costs).

The third storage function, **storage for retrieval** (labelled C), is undergoing significant changes. The traditional storage system for publications, the libraries, are supplemented by data-bases and microform retrieval systems, which store information in compacted form for mechanized and even remote retrieval. Storage for the purpose of efficient retrieval is best performed with high density storage and publication media, because high density documentation is essential for the economical automation of all retrieval operations.

Storage for information processing in support of user-system interfaces (labelled D) requires similar properties as the storage for pre-publication information processing. It calls for a memory that can be written in, erased and rewritten easily, and this with high speed. The speed requirement is unique: the storage should handle many dialogues simultaneously. Each of these dialogues between a user (in search of specific information) and the retrieval system (storing the answers to the requests) can involve extensive note-taking and information processing.

Figure 4 illustrates **past and present use of storage media and technologies for the four storage functions**.

Paper has been and still is used as the dominant storage medium for processing support. The information technology in support of input processing has been limited to the use of handwriting tools (pen and pencil) and typewriters as input systems and a wide range of paper files as storage devices. (A in Figure 4.)

The choice of publishing media was and in practice still is limited to paper, either printed or printed and bound. For some applications film media and phonograph records offer alternatives for special forms of publications. (B in Figure 4.)

Storage for retrieval was and still is implemented by organizing and arranging physically published documents for easy access. Examples range from paper files on top of desks to bookshelves and libraries housing millions of documents. (C in Figure 4.)

Likewise retrieval support has relied mostly on paper as storage medium. Indexes and catalogues written or printed on paper are the main memories in support of the user in search of stored information. (D in Figure 4.)

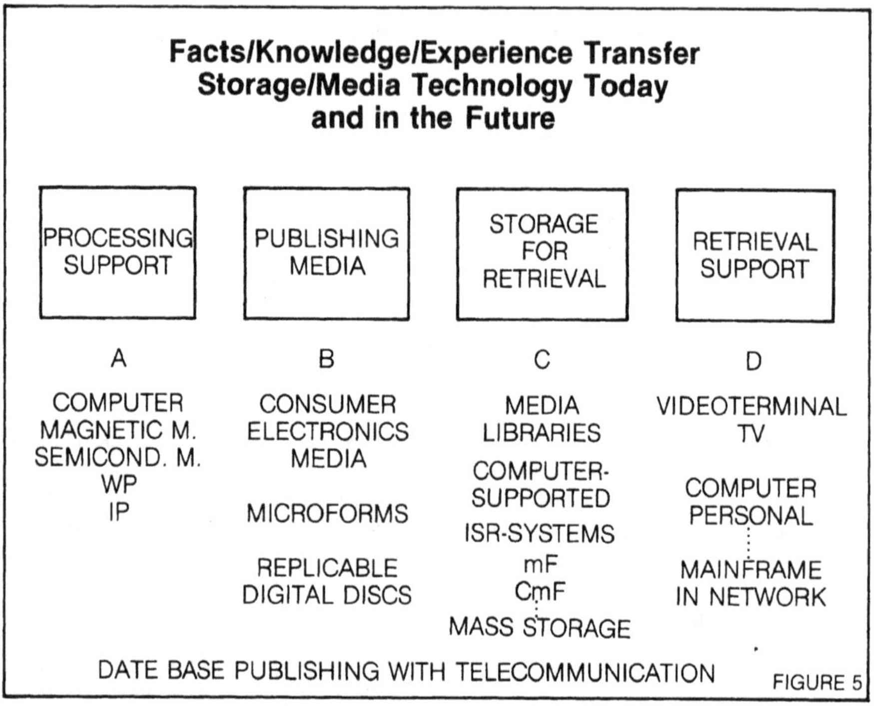

FIGURE 5

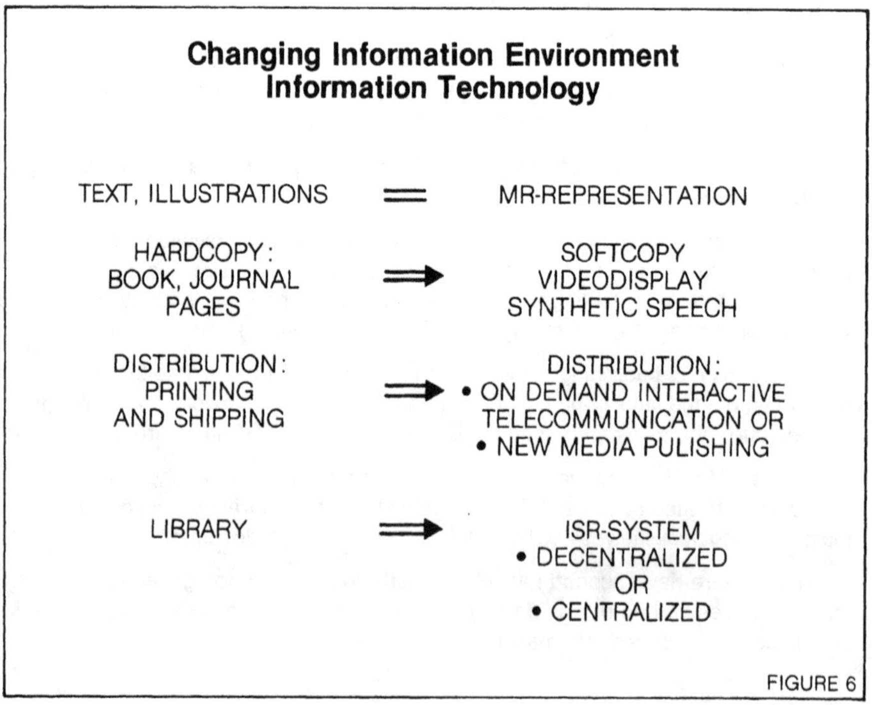

FIGURE 6

The anticipated **change in storage media and technologies** in the near and distant future is indicated in Figure 5. It can be assumed that for the transfer of fact, experience and knowledge information, paper will be practically replaced by high density MR media for all processing and retrieval support functions. For the functions of storage media and storage for retrieval it will be supplemented by various high density storage media other than paper (e.g. microforms, videomedia).

Electronic technologies and computer storage media are already invading the pre-publication processing area. Text processing and image processing systems are helping to rationalize editorial offices and speed up the print master production process. It is likely that magnetic and semiconductor memories will replace paper for practically all storage functions. (A in Figure 5.)

Paper in its function as publishing medium will maintain its important position, but it will be supplemented by new media for MR and HR data for all communication processes which involve electronic devices or telecommunication. Consumer electronics media (audio and video cassettes, software for computer games, etc.) are already used to distribute information. Microforms as publishing media for HR information can, like all high density new media, only be used in connexion with devices: readers or retrieval systems. Replicable digital discs promise to become a major publication medium of the future in the context of video and TV based home computer systems and information utility concepts like videotex. (B in Figure 5.)

Use of high density publishing media allows far-reaching changes and advances with regard to storage for retrieval. Retrieval, particularly from very large documentations in libraries, may at least be improved by computer support (on-line indexes and bibliographic data-bases) and, for specialized applications, even be automated. Automated systems are most likely information storage and retrieval systems (ISR-systems) which combine microform and digital computer technologies. (C in Figure 5.)

Information technologies in the retrieval support functions will be computer and video oriented. With the partial replacement of printed paper as the final element in the information transfer link to the user, both video and microform terminals will grow in importance. The information will be either projected on a microform screen or generated on a video screen. Computers with various magnetic media will assist in the identification, location and retrieval of documented information regardless of the type of distribution and storage. These computers may range in complexity from elementary personal computers (applied as intelligent terminals to access information networks) to mainframe computers controlling multimedia libraries and large data-bases (D in Figure 5.)

In addition, a completely new form of information distribution takes shape: electronic on-demand publishing via public telecommunication networks. It exists today as data-base publishing for retrieval support from bibliographic and numerical data-bases. Electronic publishing relies not only on easily accessible telecommunication networks, but, equally important, on the low cost of storage systems and media for MR information.

The comparison of Figures 4 and 5 emphasizes the important changes in information

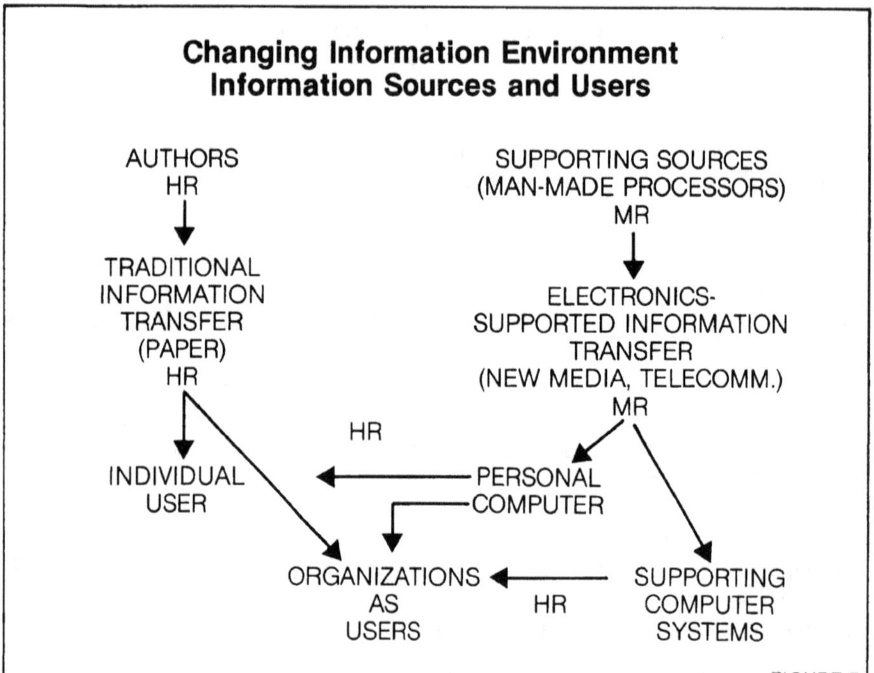

FIGURE 7

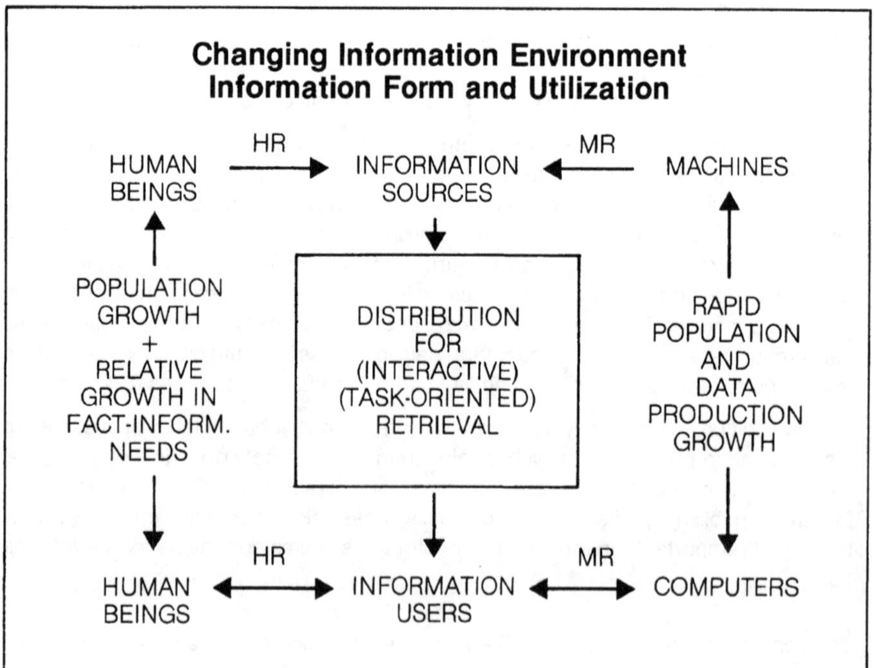

FIGURE 8

technologies and media between the past and the future:

1. Paper is for all storage functions supplemented, or, for some applications, even replaced by media with higher information densities.
2. All new information technologies are computer oriented.
3. Magnetic storage media will be used for processing and retrieval support functions.

Forces which shape the changes in information transfer methods

The predicted changes in information technologies and media are developing in response to the changing information environment (which they also help shape). The influences of three major driving forces are illustrated in Figures 6, 7 and 8.

New information technologies have led to electronic correspondences to traditional paper based concepts in information transfer. Examples are shown in Figure 6. Printed text and illustrations, forms of HR records, have equivalents in MR representations: digitally encoded code strings. The paperbound forms of display, such as book and journal pages (hard-copy), have an even wider range of electronics-based soft-copy correspondences. These include video displays and synthetic speech. Enlarged projections of microimages are another form of new soft display forms. The physical distribution methods for documents (printing and shipping) are supplemented and challenged by two different new approaches involving electronics: on-demand interactive electronic publishing (today known as data-base publishing) and publishing for electronic devices and systems with high density media for MR information. Examples are software publishing for personal computers and future publishing for display on the TV-screen with optical discs. The traditional large scale retrieval system, the library, finds its functional equivalent in ISR-systems with high density storage media. These can be decentralized. Then they depend on a supply of publications distributed with high density media such as microforms and additional MR media for the transmittal of retrieval support functions, e.g. with optical digital discs. Centralized ISR-systems are also conceivable. They store information in MR-form for on-demand retrieval and remote delivery by public telecommunication networks.

The changes in information sources and users are shown in Figure 7. Man-made supporting information sources, in particular computers and intelligent sensing devices used in research and development, are producing an increasing percentage of documented information. These supporting sources encode, unlike human sources, directly in MR-form. The information becomes the input to man-made information devices or systems: either for translation into HR displays or documents or for further processing.

New information technologies stimulate new support devices and systems for transfer of documented information. The impact of the resulting creation of new and additional sources and users is twofold:

(1) it increases rapidly the volume of documented information far beyond the population increase of man, and

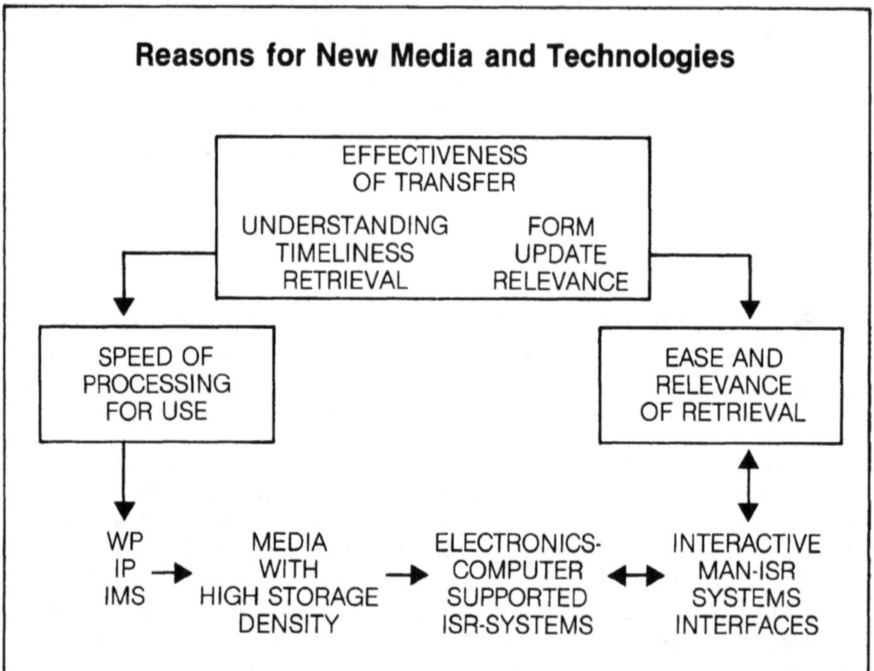

FIGURE 9

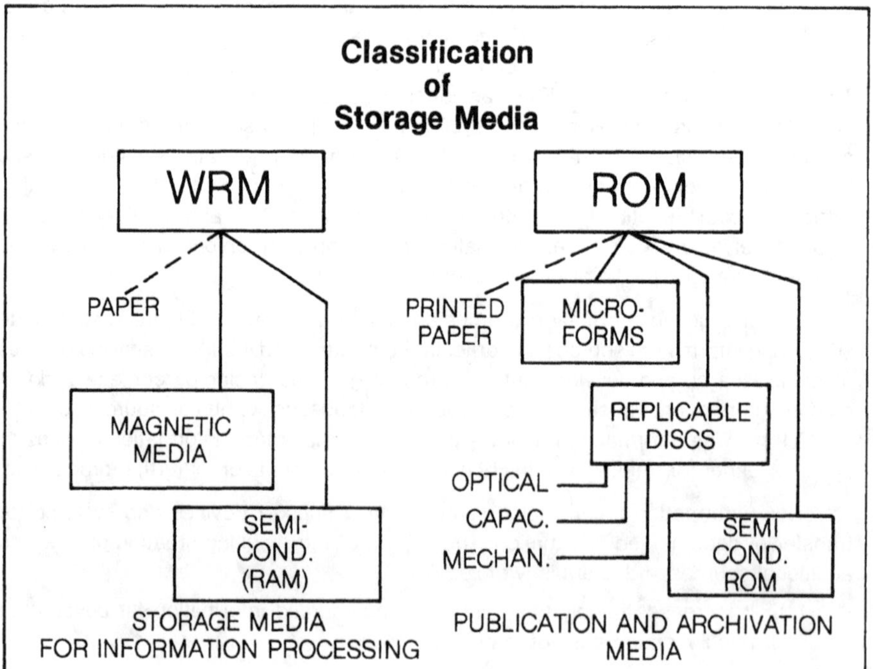

FIGURE 10

(2) it generates a completely new domain for publishing and information management, the domain of documented information in MR form.

The division of communicated and documented information into MR and HR domains is perhaps the most significant change to be faced by publishers, documentalists and librarians. The new information forms (MR) and utilizations (by machines and computers as essential support components in interactive, task-oriented retrieval functions) demand ease of convertibility from MR into HR and vice versa to allow optimal use of new media and technologies (as indicated in Figure 8). The thinking, training and planning of the traditional operators (publishers, librarians, etc.) of the fact, knowledge and experience system still centres around HR information. It largely ignores the almost revolutionary growth of MR information and its effect on user expectations and changing demands on the entire transfer systems. The deficiencies of the traditional system become more noticeable as the dependence on MR information grows: it either forces the present operators to rethink and reorient their activities to meet changing needs, or it invites the invasion of new innovative information industries into their present sphere of interest (which in the past was almost immune to invasion by outsiders).

Changing communication requirements necessitate new media and technologies

The requirements for communication of facts, experience and knowledge are changing, as the dependence on the growing population and productivity of the supporting computer system increases. Higher demands on communication speed, the need to handle MR as well as HR information and the general desire to increase the effectiveness of information systems not only favour but by necessity lead to the use of new media and electronic technologies.

There are two major driving forces in the improvement of transfer effectiveness: speed of processing for use (reduction of time between generation of new information and its availability to the user) and ease and relevance of retrieval (see Figure 9).

Processing for use involves three operations: editorial processing, distribution and organization (for retrieval on-demand), three operations presently performed in sequence by authors and publishers, the book trade and the documentation and library community. To satisfy demands for short turnaround times (arising due to the shrinking half-life spans of scientific and technical information) a tight linking of these operations appears logical. Integration of text, picture and information management processing leads to the interlinked use of computers, data communication and various MR publishing and storage media, with associated support systems.

Retrieval aims at providing the user with an easy-to-use link to the generated and stored information. In the context of computer supported systems, interactive interfaces between the user and the system are fundamental to achieving this. Digital telecommunication technologies, computers and systems using high density storage media are essentials. From both ends of the transfer link, from the information provider and from the user, the use of new media and electronics becomes a necessity: integra-

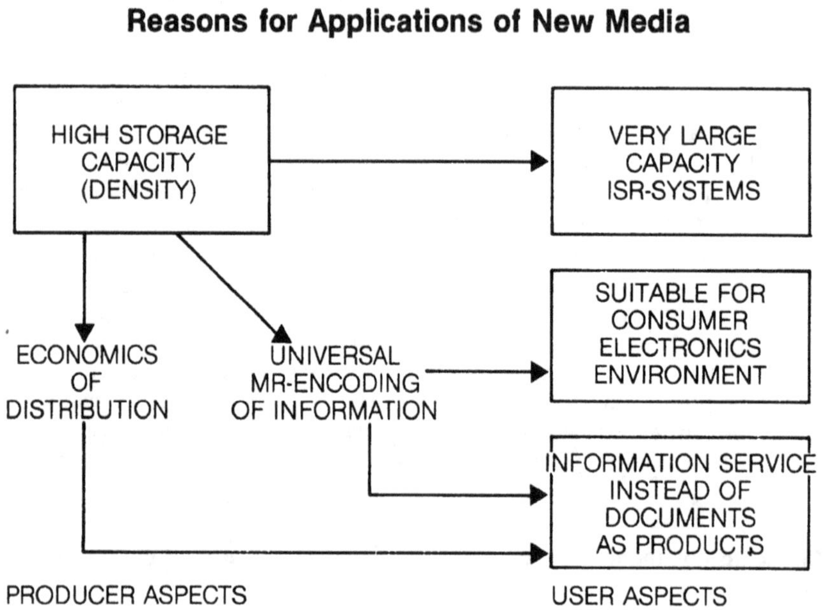

FIGURE 11

STORAGE DENSITIES

	HR	MR
PAPER	0.1 kb/CM²	
mF ~ 25x	0.1 kb/CM²	
mF ~ 50x	100 kb/CM²	
UmF ~ 200 x	1000 kb/CM²	
MAGNETIC TAPE 1600 BPI	5 kb/CM²	
MAGNETIC TAPE 6250 BPI	20 kb/CM²	
FLOPPY DISCS	n0 kb/CM²	
HARD DISCS	m00 kb/CM²	
OPTICAL DIGITAL DISCS	p000 kb/CM²	
(HOLOGRAPHIC "mF" (LAB)	p000 kb/CM²)	

FIGURE 12

ted electronics and computer supported ISR-systems promise to offer user-oriented solutions to the mounting problems associated with documentation inflation.

Storage media today: factors which determine their application

A short discussion of storage media in general is helpful in understanding the relative importance of new media. There are two significantly different classes of storage media, as shown in Figure 10. Write-read media and memories (WRM) can be used to write, read, erase and rewrite. Examples are notepaper used to make erasable pencil notes, all magnetic media and semiconductor random access memories (RAM). All these media serve storage functions as part of information processing.

The content of read-only media and memories (ROM) cannot be erased, information can only be written on once and thereafter be read unlimited times. Examples are printed paper, microforms, replicable discs (e.g. phonograph records, video discs to be used with capacitive, optical or mechanical readout devices), and semiconductor ROMs used extensively in hand calculators and computers. ROMs with their contents can be easily replicated in large quantities (large editions) and information once stored cannot be lost and is retrievable for the lifetime of the medium. These properties make ROMs the ideal media for publication and archivation.

The primary reason for the application for all new media (see Figure 11) is their high storage density in comparison to that of paper. For the information provider this can mean better economics of distribution (an example is the per page reproduction and shipping costs of microforms) and the ability to encode and store information in MR form. As pointed out before, MR information representation is essential in supplying computer support systems with data and is helpful for efficient telecommunication.

The user benefits from the high density media in various ways. They permit the design and economic operation of IRS-systems with very large capacities which allow direct interaction as part of the retrieval process. They can be compatible with consumer electronics: display on the TV-screen and data communication via the public telephone network illustrates this. The trend in user preferences for more information services, instead of documents as purchasable products, can be accommodated: the information-seeking worker, hobbyist and student is less interested in documents with a potentially heavy ballast of irrevelant information than in specific answers or comments to specific questions.

A comparison of storage densities of media may help in judging their potential. It should be pointed out, however, that storage density is only one of many parameters which influence the choice of a particular medium for a specific application: preparation, distribution and utilization costs, potential user acceptance etc. are just a few examples of these additional parameters. Figure 12 lists various existing and projected publication and storage media with order-of-magnitude statements of their storage densities. Densities are measured in kilobits (kb) per square centimetre. Media which are particularly promising as publication media are highlighted in boxes: microforms with high reduction ratios (50 times reduction or more) and optical discs with densities of several thousand kb per square centimetre.

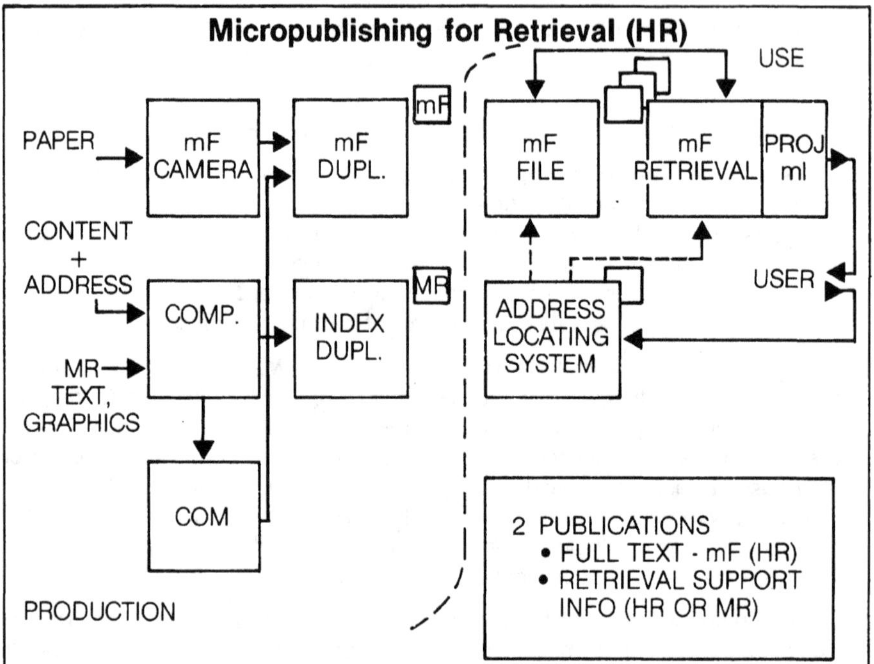

FIGURE 13

Microforms as Information Transfer Media

POSITIVE	NEGATIVE
• ECONOMICS FOR PRODUCER MASTER DUPLICATION DISTRIBUTION	• USER ACCEPTANCE
• CHOICE OF FORMATS	• READER
• HR + MR	• COLOUR ECONOMICS PERMANENCE
• EASE OF DUPLICATION	• COPYRIGHT PROTECTION
• STANDARDS	

SUITABLE FOR COMPUTER CONTROLLED ISR-SYSTEMS
- HIGH ON-LINE CAPACITY
- ECONOMIC ODP
- SPECIAL MICROFORMS

FIGURE 14

Microforms as information transfer medium

In the past, the use of microform as a publishing medium has been limited either because they were not used in efficent retrieval systems or because of inherent limitations, particularly the initial rejection by users accustomed to paper. However many specialized applications in industrial, military and business information systems demonstrate economic and operational advantages over alternative approaches. (The widely known field of micro-republishing in connexion with archivation and for file and library compaction, in which microforms replace traditional paper publications, is not primarily intended for interactive retrieval and is therefore not further discussed here.)

Micropublishing for selective retrieval rather than for linear reading and browsing should be seen as the user-oriented application of microforms in the context of changing information transfer needs. As shown in Figure 13, it involves two separate but interlinked distribution (publishing) processes: (1) publication of the full content (text and illustrations) on microforms, and (2) distribution of retrieval support information (e.g. index, access to data-base). This support information is the key to retrieval. It can be made available on any medium (paper, microform, magnetic) that provides ease of access to location codes for the full information on microforms. It can be in the form of a periodically updated catalogue or index to the microform collection. But, wherever economically justified, it is stored in MR form for on-line retrieval and update in a computer system. Micropublications as outlined here are not just collections of individual documents with one common index, but rather growing and changing sets of information which include comprehensively all available information on a particular subject field. These micropublications are resources and often essential tools for certain information workers. Their value to the user depends primarily on the human engineering of the retrieval support hardware and software and on the completeness, actuality and relevance of the information that can be obtained in response to a particular problem.

The production process for a micropublication reflects this dual information flow. The content is either microfilmed (from printed text and illustrations as originals), or, if available in MR form, converted directly into microfiche by a COM-system. This primary information is inexpensively duplicated and distributed as duplicate microforms. In a parallel process the secondary information is generated, usually with a computer supported system: keywords with associated microimage addresses. These are sorted and organized in a data-base management system. The index information is therefore always in MR form. This permits changes and updates economically and encourages the further use of index information in a computer and telecommunication environment (e.g. in an on-line search system).

In micropublications which are used in computer assisted systems, the user has no direct access and contact with the printed page. Instead, interactive search in a data-base allows fast identification of relevant items in terms of addresses. Physical retrieval of the microimages from a microform file may be achieved with modest means manually or with sophisticated automated systems under the control of the search system. The display is either by projection of the microimages onto a screen, or by on-demand full size reproduction on paper. Micropublishing for retrieval is in essence a continuously updated publishing of centrally managed specialized libraries.

Microforms permit (as publication and storage media) compaction, organization for ease of use and partial or full automation of retrieval. Most important however are the economic advantages and the potential for up-to-date complete information supply in a decentralized and user-oriented form.

Microforms as an information transfer medium exhibit many desirable features but are of course not without disadvantages. In Figure 14 positive and negative attributes of microforms are listed. Advantages are evident for the information provider: production costs for master and duplicates are low and do not depend much on the number of copies made, distribution costs are modest because of low weight, and international standards favour worldwide marketability. Although the outside dimensions of microforms are standard (postcard size A6 format for microfiche), the size of microphotographed originals can range from that of a catalogue card to that of large maps. Microforms usually are for HR information, but they can also carry MR data - in its simplest form bar codes or bit patterns which can be optically sensed with special micro-readers.

The negative properties of microforms are of no direct concern to the information provider, but they are most noticeable to the user. Dependence on a reader and a power source and the lack of colour (it is available but is relatively expensive and can fade when frequently projected) are often disliked. Inability to make notes and to browse simultaneously in two or more documents are additional reasons for only reluctant acceptance by the user.

Ease of duplication (microfiche duplicators are becoming common office machines like office copiers - in one operation a microfiche with up to 400 pages can be duplicated at the cost of a single photocopied sheet) worries the copyright owner since no effective protection of his rights besides enforcement of the copyright laws is known.

Microforms are suitable for use in computer controlled IRS-systems with automated retrieval and on-demand reproduction. These computer microform (CMF) systems offer high on-line storage capacity and economic on-demand publishing capabilities, provided special microforms with very high reduction ratios (e.g. ultrafiche) are used.

Optical digital disc as an information transfer medium

Digital optical discs are an outgrowth of the videodiscs which are presently being developed worldwide by major electronics companies. They promise to become perhaps the most important new publishing medium in the decades to come. They offer most of the advantages of microforms without their disadvantages and numerous additional advantages for operation in a consumer electronics environment.

Commercial introduction of this publishing and storage medium awaits agreement among the major producers on a generally acceptable standard for physical properties, formats and data layout. Most technical problems are solved and prototypes of various systems have been demonstrated in the USA, in Europe and in Japan. Its appeal as a storage medium for home and business computers may accelerate its

Disc-Publishing for Retrieval (MR)

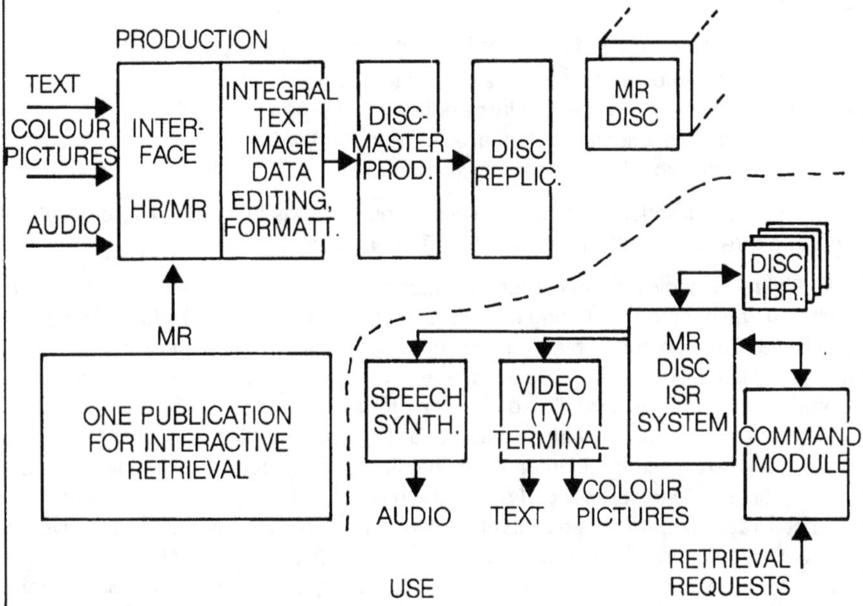

FIGURE 15

Optical Digital Discs as Information Transfer Media

POSITIVE	NEGATIVE
1 ECONOMICS OF SCALE DUPLICATION, HARDWARE	1 LIMITATIONS IN DISPLAY FORMAT
2 UNIVERSALITY. MR: FOR TEXT, COLOUR PICTURES AUDIO, COMPUTER DATA	2 EDITORIAL AND MASTER PRODUCTION COSTS
3 POTENTIAL COMPATIBILITY WITH PERSONAL COMPUTERS, TV, TELECOMM.	3 COST OF RETRIEVAL SYSTEM
4 HIGHEST PRACTICAL INFORMATION DENSITY FOR ECONOMICAL PRODUCTION TECHNOLOGIES	

FIGURE 16

commercial development for introduction as part of computer subsystems in the mid-eighties.

Future optical digital discs offer three major attractions:

1. the highest economically realizable storage density.
2. very low publication costs (if large editions are produced), and
3. the universal application potential as carrier of MR information for text, colour pictures, audio signals and data for computers and control devices.

This last attraction permits the design of publications for interactive retrieval with one single record for visual, audio and MR information.

Figure 15 shows the major functional components, which when combined represent retrieval oriented publishing with discs. All forms of HR and MR information can be published. Text, colour or black-and-white pictures as well as audio signals are first translated into MR code. Text, colour image and audio signals are merged in integrated editorial and layout systems. The data is then formatted and preprocessed for automatic disc master production with a (laser) recording system which is functionally comparable to a video cassette recorder. From the master disc a replica is "printed" in a process analogous to the pressing of phonograph records. MR discs can carry (per 30 cm diameter record) up to the equivalent of more than one million printed pages (average content 1 000 characters) for display on TV-screens together with associated retrieval support information as needed to search and locate specific items in this single record library.

Information published with digital optical discs can only be accessed and made readable with compatible IRS-systems. These can be elementary home systems consisting of a personal computer as command module, a digital disc player similar in size and cost to video players already available for video discs and a TV-set for display. Manual exchange of discs out of the disc library allows playback and retrieval of whatever may be published by this new medium: educational programs, computer systems and application software, publications formerly distributed in paper or microforms including fully organized, indexed and periodically updated libraries. The information may be displayed either as a colour image on the screen or as sound. Speech can be synthesized digitally from the digitally encoded phonetic text, or complex sounds like music and background in a concert hall can be reproduced from digital sound tracks. More sophisticated systems are conceivable, for installation in public libraries, with access by many users, and partially automated retrieval, loading and filing of selected discs from and in the library. Discs may also become directly accessible by telecommunication from home or office against a metered fee.

This futuristic and optimistic portrayal of potential disc-based information transfer and publishing system ignores the negative factors which may delay or even prevent the digital disc from attaining a dominant position in the information world. Pro and con characteristics of optical digital discs are summarized in Figure 16. The most likely acceptance delaying factor will be the cost of the retrieval system to the user. Other negative factors are the high expenses for the preparation of material for recording, the costly specialized generation of information for interactive use (comparable in cost to

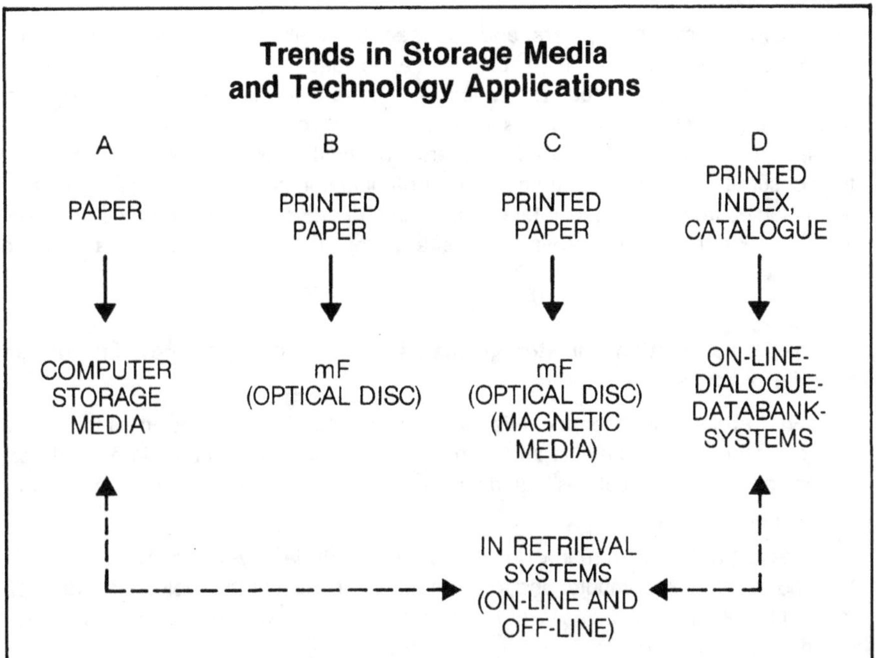

FIGURE 17

Changes in Information Transfer Practices made possible by New Storage Media/Technology

- SHIFT IN RATIO FORMAL/INFORMAL COMMUNICATION
- GROWING SIGNIFICANCE: MR-PUBLISHING
- NEW PUBLISHING OBJECTIVES
  - FOR RETRIEVAL SYSTEMS → HIGH DENSITY MEDIA
  - FOR INTERACTIVE USE → MR-ENCODING
  - FOR COMPUTER-BASE SYSTEMS MR-DATA BANKS
- DISPLAY FORMATS DETERMINED BY CONSTRAINTS INHERENT IN TECHNOLOGY

IMPACT LIMITED TO INFORMATION TRANSFER OF FACT, KNOWLEDGE (ENTERTAINMENT) AND EDUCATION-ORIENTED INFORMATION PRODUCTS

FIGURE 18

that of programming computers) and the need to reformat all information to become compatible with the restricted display capabilities of consumer video devices (TV-technology). Positive features which have been discussed already outweight the negative ones, at least when analysed from a rational point of view. The most important ones, for practical reasons, are the universal application both in the computer and in the consumer electronic environment, the potential compatibility with well established technologies (television, public telecommunication and computers) and the promising improvements in turn-around time and effectiveness of the information transfer process.

Trends in applications of storage media and technologies: new information transfer practice

Four areas have been identified in which storage is important. In all four areas new computer oriented media and technologies are challenging and, in part, already replacing traditional practices based on paper. These major trends are summarized in Figure 17.

These transitions from low density paper to high density media, MR documentation and reliance on computer and telecommunication favour the evolutionary development of far-reaching changes in information practices. A list of the major ones appears in Figure 18.

Public access to worldwide telecommunication may be the major contributor to a change of the ratio between formal and informal communication in favour of the latter. This can reduce the need for traditional forms of publishing in some areas, but is unlikely to influence the requirement to document fact, experience and knowledge information. The practice of documentation may be different: automatic recordings of dialogues may replace in part memos and manuscripts. Recordings may be translated into MR form for ease of universal future use. This may add to the growing significance of MR publishing, which is already evolving outside the traditional publishing community: within the computer industry it exists already to satisfy rapidly growing software needs.

New and additional publishing objectives are emerging: high density media permit publishing for retrieval systems, MR encoding changes the content of publications from monologue style to an engaging interactive dialogue style. Field , work, profession and problem oriented MR databanks are likely to provide new markets for the information industry. These databanks require updating and continuous maintenance. They can provide steady sources of income.

Concluding comment

It is very likely that the impact of the new media and storage technologies will be limited primarily to one segment of publishing: that dealing with scientific, technological, social and economic matters and the transfer of facts, experiences and knowledge. It should be noted however that this segment is of great significance for the future of mankind.

Teleordering in Denmark

Bjarne HJORT

Gyldendal,
Copenhagen,
Denmark

The Danish bookshop terminal system (Danske Boghandleres Terminalsystem) is an information and order retrieval system which began in 1974 as an internal system in the Gyldendal publishing house, to help the processing of written orders from booksellers. It was soon suggested that terminals be placed in bookshops, so that the entering of orders could be done by the booksellers themselves. Four bookshops participated in an experiment in September 1975. This was a success and led us to decide that we wanted to see established in Denmark a single teleordering and book information system covering as many publishers and bookshops as possible.

We spent all of 1976 on discussions with other publishers and on assessing interest in the system among booksellers. Eventually agreements were reached which resulted in about half of all the titles published in Denmark by various publishing houses being included in the system, and 110 bookshops, which between them were selling about 45% of all books sold in Denmark, decided to participate. Installation began in September 1977 and by the end of 1977 the system was running.

Since then more publishing houses have joined. Bearing in mind the frequently-made comment that the new technologies are going to kill off the small publisher, it is interesting to note that among the participants are a number of small publishing houses with lists made up of as few as five titles. At present 85% of all titles published in Denmark are included in the system and we are expecting this figure to have reached 95% by the end of 1980.

As well as its use for ordering, the system acts as a computer catalogue, listing all the books which are currently available from the participating publishers and also books which went out-of-stock during the previous year.

There are two ways of seeking information in the system. The quickest involves entering 2-4 characters of the author's name and 2-4 characters from the title of the book, and thus relies on the title and name of the author being known to the enquirer. This method gives a unique answer to 80% of the enquiries. An order for the book thus identified can then be placed through the terminal, without the bookseller having to concern himself about which publisher or distribution centre will be responsible for sending the book to him.

Before the introduction of the system, the bookseller would have dealt with a request for a book issued by a publisher whose identity was not known by searching for the book in his collection of publishers' catalogues. He would have begun with the catalogues of the major publishers, and if the book was not to be found there he would probably have given up the search rather than spend his time working through the catalogues of the small publishers. This unconscious bias against the smaller publisher is of course avoided in the new system, where titles from small publishing houses can be identified just as easily as those from the largest firms.

A second method of seeking information in the system is less restrictive than the first. The enquirer may enter as much as he can recall about the author and title of the book in question, using from one to ten words, and the system will attempt to correct any spelling mistakes.

Orders can be placed by the bookseller in two different ways. He can direct the terminal to transmit an order for immediate fulfilment, or he can arrange for the order to be "parked" in the computer for as long as the bookseller desires. This latter facility allows the bookseller to take advantage of arrangements under which publishers offer him a higher discount if he purchases a certain quantity of books at the same time, rather than a single copy.

Orders sent in by bookshops are extracted from the central computer and sent to publishing houses or distribution centres twice per day, either printed on paper or as magnetic tape to be processed by the computer system of the publisher or distributor.

Right from the beginning we decided that the cost of the system should be shared between the publishers and the booksellers. To become connected to the system the bookseller has to invest £2 200 for the terminal, modem, telephone lines, and his share of a small computer which we call a concentrator. Thus the system up to the front of the computer is owned by the bookshops. The bookseller also has to make a monthly payment of £110, for service to the terminal and concentrator and to cover payments to the telecommunication authorities for the modem and the telephone line. All bookshops pay the same amount, whatever their size or distance from the computer. It should also be noted that we use leased lines, so that the terminal is open all the time - it is not necessary first to "dial-up" on the telephone. The average line length is 24 km, which is low and arises because we are using a multidrop net with concentrators out in the country rather than immediately in front of the computer.

The computer is paid for by the publishers. When a publisher's whole stock is first being entered there is a charge of £2.25 per title. Later there is a charge of £4.50 for each new title added. Corrections to the data-base, or catalogue, are free however, because we want to encourage publishers to keep it as correct and as up-to-date as possible. The publishers also pay a charge for stocking titles in the system, which is £1.50 per title per year. Each order line which they receive is charged at 2.5 pence. No charge is made if a publisher wants to connect a terminal to the system so that he can see for himself what information has been entered.

A special effort has been made to create a system which is easy to use, and we have found that only one hour is needed for all the staff of a bookshop to be taught how to use it.

By the end of 1980 we expect that about 250 bookshops will be involved and that about 75% of all sales of books in Denmark will be through the system. At present the system is receiving 40-50 000 questions per day and 8-10 000 order lines. The average question contains 8 to 10 characters and the answers to the bookshop 175 to 200 characters on average. Some bookshops are using their terminal as many as 7-800 times per day, which of course with a dial-up system would be impossible. The system is being used both by very big bookshops and by small bookshops, the latter sometimes run by a married couple with no assistance, who find that the terminal saves them time in the running of the bookshop for a monthly outlay (£110) which is less than would be needed to employ an assistant.

Our plans for 1980 include the development of the system so that it can help with bookshop accounting, with the operation of bookclubs (which in Denmark are being run by the bookshops), and with mailing systems. We also intend to introduce an order reminder facility, which will enable booksellers to see what books they have on order.

About 40 bookshops have connected a printer to their terminal and a further 40 bookshops have printers on order. These bookshops will thus be able to extract information from the system in the form of hard copy, and we intend to make use of this possiblity by putting the dictionaries and encyclopedias of our publishing house into the system. Other publishing houses will be able to do the same, and the bookseller will then be able to provide a new service: his customers will be able to go to him to buy required extracts from, say, an encyclopedia, printed on paper.

In conclusion, I would say that we are not afraid of the new technologies. On the contrary, we are making use of them, but in a controlled way as a supplement to the printed book, rather than as a replacement.

Prestel

A.D.W. STEPHENS

Head of Prestel International Division,
Prestel Headquarters,
Post Office Telecommunications,
London, United Kingdom

"Prestel" is the name given to the "viewdata" service of the British Post Office. We would describe viewdata as a system which converts a suitably modified domestic television set into a terminal for use in the home or a business environment, the television set being connected to one or more remote computers, on demand, using the normal public telephone or similar network. It enables words, figures, diagrams, histograms and simple pictures to be displayed on the television screen interactively following commands from an input device, which maybe a keypad looking not unlike a pocket calculator or from a full alphanumeric keyboard. The size of the display is 24 lines of 40 characters (this varies slightly according to the specific implementation) and is therefore roughly equivalent to one quarter of the capacity of an A4 sheet.

In the United Kingdom we have been using "viewdata" as a word to cover services exemplified by Prestel, but it has caused problems in that "viewdata" cannot be readily pronounced in some languages. The word "videotex" has therefore been tentatively adopted internationally, and can be divided into "broadcast" videotex which covers services such as "Ceefax" and "Oracle" established by the broadcasting authorities, and "interactive" videotex, which is synonymous with viewdata. Viewdata and videotex are thus generic terms: in the United Kingdom "Prestel" is the name given to the specific service of the British Post Office. Analogous terms for similar services run in France and the Federal Republic of Germany are, respectively, "Teletel" and "Bildschirmtext". Other names have been coined for systems being developed in other countries.

One of the special points of the Prestel system is that all the extra electronics required to make the television receiver operate as a terminal - the modem, autodialler, storage, decoder - are contained within the modified receiver. The result is that it looks like a normal television set, except that there is an extra cable coming out of the back, which is plugged into the telephone network. We consider this to be a good sales feature of the Prestel system: some other administrations, because of regulatory problems, are not able to use integral modèms, and thus in some systems there has to be an external box connected to the television set.

The Prestel data-base is operated by the Post Office on a common carrier basis, which means that the Post Office provides the computers and storage into which anyone willing to pay the Post Office's charges can insert information, and from which others can extract information. We exercise no editorial control whatever over what is put into Prestel. We do of course offer advice, and in certain circumstances, subject to the rule of law, we may have to expunge information from the data-base. So far this has been done on the grounds of obscenity only, although there could be other reasons which however have not yet arisen.

At present the capacity of the data-base is 250 000 pages, most of which have been sold to information providers. Capacity will be increased to 500 000 during 1980. The service is currently available in London, and is being expanded over the next twelve months to 17 other towns and cities around the United Kingdom.

It is hard to give specific guidance to publishers, authors and booksellers about the relevance of Prestel to the publishing industry, except to say that the viewdata type of service is a new medium waiting to be fully exploited. Very few people have yet got firmly to grips with it. So far we have about 170 information providers, supplying information in such disparate areas as timetables (air, rail and sea), financial affairs (shares, commodity prices etc.), programmed learning, news and sport, and advice to householders (first aid, gardening, decoration etc.). There is a facility for sending messages to information providers through the Prestel system, and advertising is included with the possibility of ordering goods through the system. Payment is made by sending one's credit card number, through Prestel, to the vendor, and the goods available include alcoholic beverages, cameras and household appliances.

Viewdata is not a medium which traditional publishers should be frightened of, nor should it be regarded as directly competitive with traditional publishing methods. The two technologies can and will coexist and this is best demonstrated by the fact that approximately half of the information providers on Prestel are involved in traditional publishing businesses.

A few of the possibilities of Prestel are shown in the following Prestel "pages" photographed on a television screen. Up to seven colours are used in the Prestel system, which are of course lost in these black-and-white reproductions.

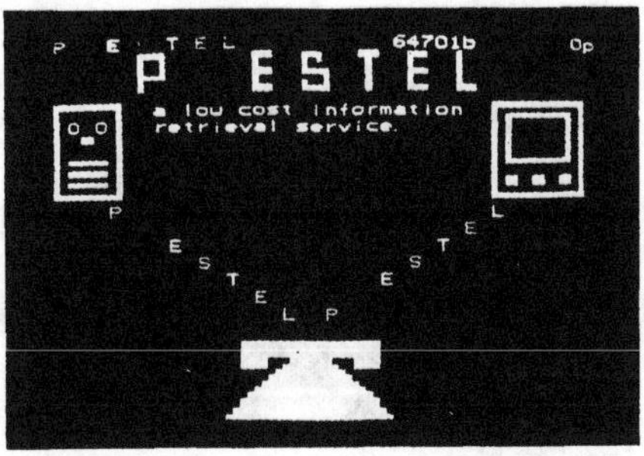

Figure 1.
A simple diagram showing how the Prestel service links an adapted television set, via the ordinary telephone line, to a vast computer-held store of information which is relevant to both business and private users.

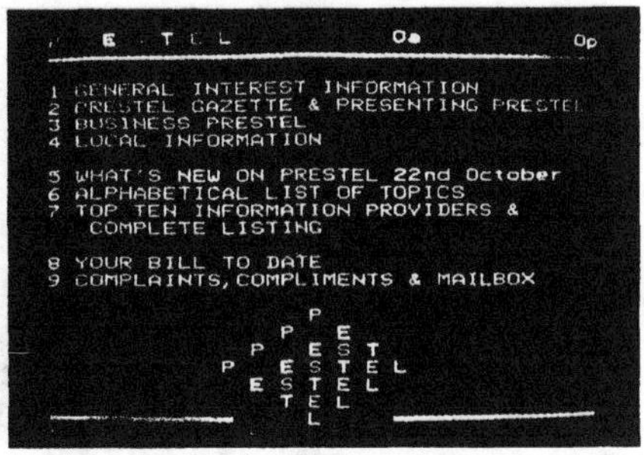

Figure 2.
The basic index page.

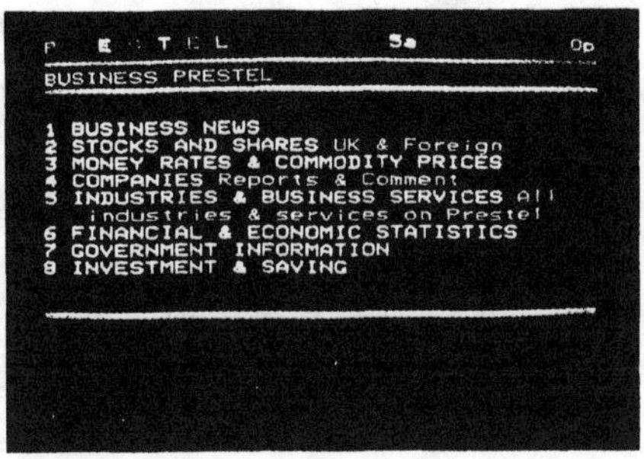

Figure 3.
The main business index page gives an idea of the range of information available.

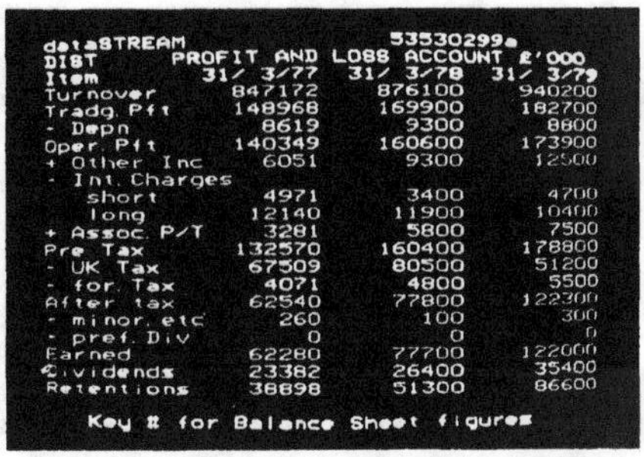

Figure 4.
This page is an example of the company information which Prestel can supply. Information is available for approximately 1 000 firms.

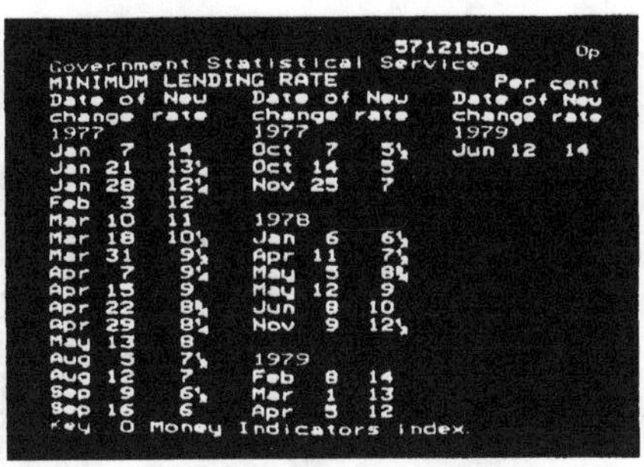

Figure 5.
Movements in Minimum Lending Rate are dealt with on this page.

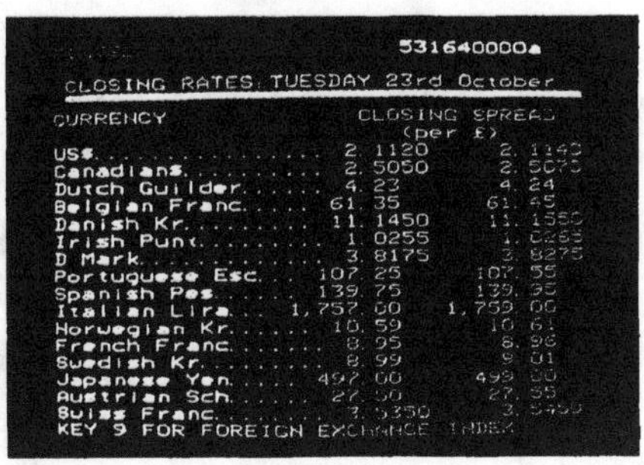

Figure 6.
Up-to-date exchange rates are also listed.

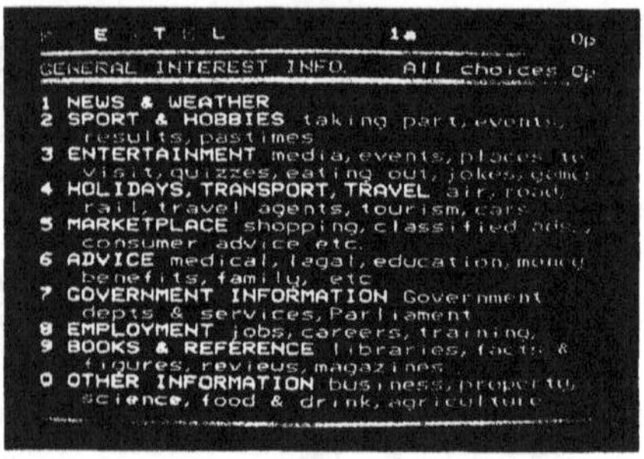

Figure 7.
This index page shows that Prestel contains plenty of information to interest the user at home.

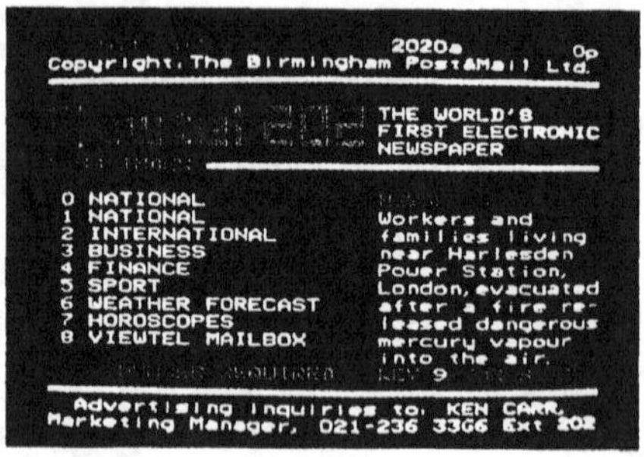

Figure 8.
The latest news is readily available.

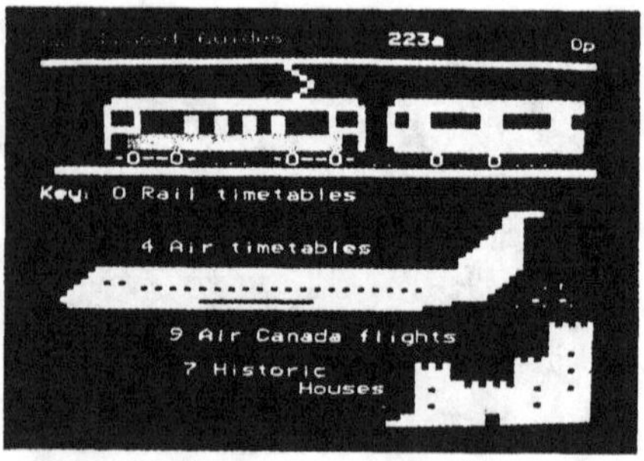

Figure 9.
This page is an introduction to travel information on Prestel and shows the kind of pictures which can be used to enliven the page.

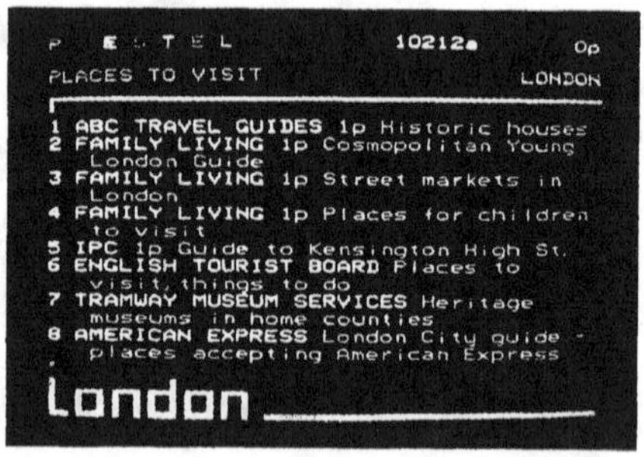

Figure 10.
Prestel will supply ideas on what to do on a day out - a basic index page.

Figure 11.
Places to Visit in London - an index to the offerings of one of the information providers.

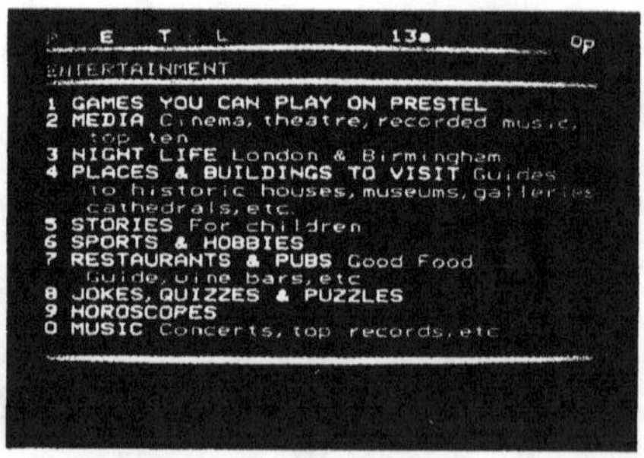

Figure 12.
Entertainment - an index page.

Figure 13.
Games can be played on Prestel too.

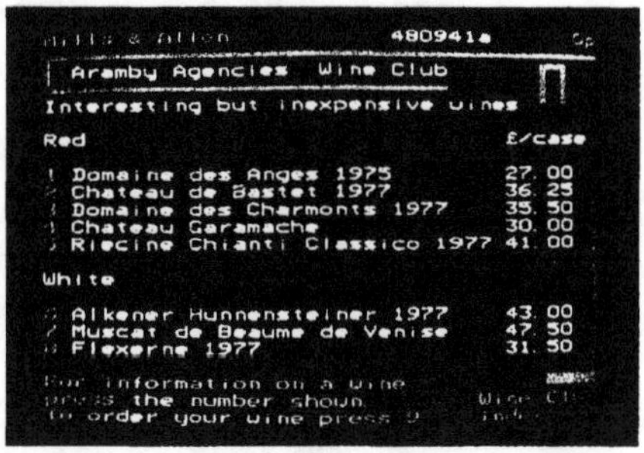

Figure 14.
By using their credit cards, users can place orders through Prestel for a wide range of goods.

French policy on videotex

Hervé NORA

Head of the Telematics Division,
Directorate-General for Telecommunications, France

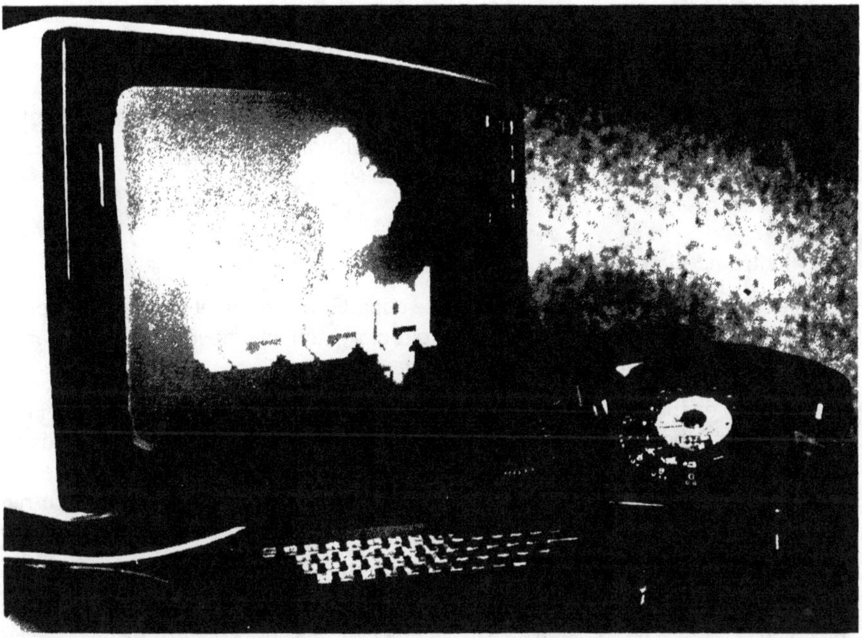

What is videotex?

Videotex is the use of a television screen to display data, information, alphanumeric or graphic messages.

Videotex may be interactive as in the case of TELETEL, or broadcast as in the case of the ANTIOPE service of TDF (Télé-Diffusion française). The French standardization proposal (ANTIOPE standard) ensures compatibility between the interactive and broadcast videotex systems by simply incorporating a "DIDON" card into the interactive videotex terminal. This terminal may take two forms:

a unit, comprising decoder, modem and alphanumeric keypad, connecting the domestic television receiver to the telephone network; this device may or may not be built into the television set;

a special terminal in which the television screen is fully-integrated with the decoder, modem, keypad and possibly telephone; the 'electronic directory' is an example of this.

There are various reasons for developing a fully-integrated terminal:

the telephone and television set are usually located in different rooms;

the fact of having only one screen for watching television programmes and, for example, consulting the electronic directory would raise quite a number of problems;

the cost of a mass-produced electronic directory with a white screen of approximately 20 cm should work out at a fairly moderate figure, in the region of FF 300 to FF 400, excluding tax.

Whatever its form, the videotex terminal will be a new means of communication opening up the possibility of complex dialogues with other terminals and data processing services of all kinds, in particular data banks.

The electronic directory

In 1982, the 250 000 subscribers in the département of Ille et Vilaine will be provided with the special videotex 'electronic directory' terminal described above, which will be installed in their homes free of charge.

This will enable them to look up the names of telephone subscribers (initially in Ille et Vilaine and subsequently all subscribers in France). They will also be able to consult the professional directory and, if they so wish (that is at their request), have access to additional information of an advertising nature. This information will replace the current advertisements in the printed directory. However, the exact form has yet to be decided.

The electronic directory should improve significantly the 'directory inquiries' service, particularly since the directory file will be updated weekly and possibly daily at a later stage. There is now no doubt as to the technical feasibility of the project; however, various obstacles have yet to be overcome, particularly in relation to the 'man-machine dialogue': it is essential that the service can be used without training *at least* by all those who now know how to consult the printed directory. Here, the design and ergonomics of the keypad will undoubtedly be important factors.

If the Ille et Vilaine experiment proves successful, the service will gradually be extended to the whole of France over a period of some 10 years: in 1992, approximately 30 million electronic directories should be in service. This development will be financed partly by savings on the printed directory (estimated cost in 1983: FF 50 per annum per subscriber) and by charging inquiries to the directory at the rate for local telephone calls.

This slight increase should be acceptable to the subscriber in view of the substantial improvement in the service provided: for an average subscriber who consults the printed directory once a week, the cost would amount to FF 25 a year at current tariffs.

Electronic directory TRT (Télécommunications radioélectriques et téléphoniques)
(Photograph Télécommunications/SCORE)

The Vélizy experiment (TELETEL project)

The range of services (other than the electronic directory) which can be accessed via a videotex terminal will be tested in the Vélizy area as from the end of 1980 (communes of Vélizy, Versailles, Buc, Jouy-en-Josas, Les Loges-en-Josas).

This experiment, which will cover 2 000 to 2 500 households, is designed to test the services, to assess the degree of interest they arouse among the general public and their acceptability, and above all, to *examine the economic and social consequences of each of them on a realistic scale.*

Such a study is required since it is impossible to examine 'on paper' the effects of the services on the various economic sectors and whether or not they are compatible with existing regulations, particularly on information, competition and advertising.

Someone with a videotex terminal who wants to go to Japan will not call up the 'travel' heading of a centralized data-bank: he will call Air France, Jet Tour, or more likely one of his local travel agencies. The latter, like most small service organizations, will probably advertise its services through an intermediary, for example a newspaper.

To take a concrete example, it can be assumed that for the purposes of information and advertising, the travel agents of Vélizy and Versailles usually contact their regular customers through the local paper, the *Nouvelles de Versailles*. The subscriber will therefore consult the 'travel' section of the service offered by this paper, which could also provide facilities for personal advertisements, local infomation, etc.

A data-processing centre, managed by the Directorate-General for Telecommunications, will be available to potential providers of services for the duration of the experiment to allow them to participate in the Vélizy project. Use of the centre will be optional and it will be disbanded at the end of the experiment.

The charges for participation in the experiment will be as follows:

the subscriber will be charged for his communications with the TELETEL centre by time units (probably a basic charge every five minutes) and the service itself invoiced at a price determined by the supplier of the information; this service may be free of charge;

the service provider who uses the Vélizy data-processing centre will be invoiced with a charge based on the machine time he requires;

the provider who does not use the Vélizy centre will be accessible via the TRANSPAC network, the corresponding tariff being applied.

The number of firms or organizations planning to provide services in Vélizy will probably reach 200 by the end of this year. A few examples:

the SNCF (French Railways) will provide a seat reservation service, customers having the option of collecting their tickets at a station or travel agency or receiving them through the post;

several banks will offer their customers a service allowing them to consult, in their homes, their bank statements showing the last ten entries, and a service for paying bills by telephone; this service will require the incorporation of an identity card reader (probably "memory" card);

several mail order firms will offer services for automatic ordering and payment, the latter in cooperation with the banks referred to above;

in addition, through the CEESI (Information Systems Study and Experimental Centre), which is responsible for coordination under the Secretariat General of the Government, all the central government administrations and other organizations in the public and semi-public sectors will provide users with basic administrative information (more than 20 000 pages of current regulations, rights and procedures, etc.) and a message service linking up the interministerial administrative information centres and the main external services concerned. Subsequently, it should be possible to develop services making it possible to complete administrative formalities from the home or from various public places;

the Directorate-General for Telecommunications will offer an information service, particularly on connexions and relevant charges, and for the payment of telephone bills; it will also offer a user advisory service and a directory of the organizations providing services.

In addition, 1 000 of the 3 000 terminals used in the experiment will give access to ANTIOPE services offered by the television channels and broadcast by the TDF.

The planning and conducting of this experiment will be carried out in close cooperation with all the partners involved, the local elected representatives and, of course, the population concerned. A permanent press-telecommunications working party was therefore set up in June 1979 to study the problems raised by this new venture.

After Vélizy

Following the Vélizy experiment, that is towards the end of 1982, the experience gained will be used to decide which services should be supplied to subscribers with videotex terminals (i.e. subscribers with the electronic directory or who have bought their own directory terminal, a television set adapted for videotex or a terminal allowing them to use their television set). Conditions for using the services supplied will also be laid down. Any amendments required in the current laws and regulations will also be made. The services will then be offered:

directly by firms or public bodies with their own data-processing hardware (this could amount to no more than a personal computer, for example, in the case of a tradesman);

through data-processing services companies, for firms not wishing or unable to install hardware of their own.

Current thinking suggests that local services will be accessed directly via the local telephone network. For obvious reasons relating to transmission costs, it will probably be necessary to go through the TRANSPAC network to gain access from, for example, Limoges to the SNCF reservation service located in Paris.

It may therefore be necessary to set up specialized access points which will form a TELETEL network, although it is possible that the future, bottom-of-the-range switching centres will be able to fulfil this role.

The services will be listed in the electronic directory of telephone subscribers (detailed arrangements having yet to be defined).

The subscriber will be invoiced:

by the Directorate-General for Telecommunications, for the cost of his telephone calls;

by the supplier of the service, for the cost (if any) of the service;

it can therefore be anticipated that specialized invoicing companies will be set up, unless the use of electronic payment instruments and in particular "memory cards" can provide an appropriate solution to this difficult problem.

Professional videotex

As soon as videotex terminals for the general public are available on the market in sufficient numbers, that is as from the end of 1980, public bodies and private firms will be able to set up internal information networks or extend their existing networks (telephone and data transmission networks) to which these terminals or similar versions adapted to the requirements of specific professions will be connected.

Various applications, currently still limited due to the cost of terminals, could be made available on a given geographical site or shared by various establishments. Such applications include:

consultation files;
management (stocks, orders, deliveries, invoicing, accounting, etc.);
internal exchanges of electronic messages.

There are already a large number of firms working on such professional, i.e. job-oriented, applications of videotex.

Télétel: Vélizy 1981
(Photograph Télécommunications/SCORE)

Euronet DIANE, the European network for direct access to scientific, technical, economic and social information

S. LUSTAC

Directorate-General for Scientific and Technical Information
and Information Management,
Commission of the European Communities, Luxembourg

Abstract

The combination of informatics and telecommunications permits the development of new methods of direct information access which can have far-reaching implications for the traditional world of publishing. The systems, aimed at the professional user rather than the general public, are characterized by the power and sophistication of their command language. Euronet DIANE is an example. Using the Euronet telecommunications network, this European Community system provides cheap and easy access to 23 DIANE hosts offering on-line access to some 175 data bases. Far from being a threat to publishers, Euronet DIANE can open up new prospects by stimulating the demand for primary documents and helping publishers to become data base producers.

Introduction

In less than thirty years, publishing and the press have experienced two revolutions. The first of these was the explosive development of audiovisual media, particularly radio and television. The second, which is likewise based on technological progress and the fall in cost of electronic components, is now under way. It is characterized by the combination of telecommunications and informatics and, more particularly, by the newly available possibility of interrogating a computer from one's home or office for the purpose of retrieving the references of a document or specific work and in certain cases the actual content of the document in question.

This prodigious technological development may well frighten those who have not yet mastered computer technology, because when telecommunications and informatics are combined and available on network, not even its most determined opponents can avoid it. The aim of this paper is to allay their fears by explaining, with Euronet DIANE as our example, the nature of the new applications now emerging in the field of direct access to information and to demonstrate that there is no need to be an information expert to profit from these new technologies as a user or supplier of on-line services.

The elements of a network providing direct access to information

All information, whatever its medium, must be collected, processed, stored and distributed. Thus, for example, a manuscript is written, edited, printed and subsequently distributed in book form to booksellers who offer it for sale to readers. An on-line information network carries out basically the same functions; but the information, instead of being stored on paper, is stored on magnetic tapes or discs, and instead of distribution by train or truck, the network uses telephone lines to reach the intermediaries or if so desired the users, direct. The two distinct functions which correspond to information collection and distribution are those of the data-base producer and the host.

a) The data-base producer:

The producer collects the unprocessed information for storage in a form usable by the hosts' computers. There are all kinds of data-base producers. They may be firms, professional associations, universities or administrative departments. In many cases, this function is first approached as a secondary activity exploiting the by-products of the main activity. Thus, the first bibliographic data-bases were compiled from the tapes used for the photocomposition of abstracts journals or indexes. But although the production of data-bases can be the secondary activity of the traditional operators in the information market, there do exist, nevertheless, many professional producers specializing in the making of this type of product even if their income is derived from a whole range of sources.

It is a convention to make a distinction between bibliographic data-bases which provide references to primary documents and factual or numerical data-banks which provide direct access to the information itself. In this paper we shall frequently use the blanket term "data-base" to cover these two aspects.

b) The host or information distributor:

The host's function is that of commercial exploitation of direct on-line access to data-bases. His role is therefore clearly more complex than that of the producer.

He has to analyze the information market both upstream and downstream, upstream to find the information resources the access facilities to which he will commercialize and downstream to find out who is going to use his services.

He must possess adequate computer capacity to store his files, and also to convert the non-computerized documentation collections of traditional holders of information.

Above all, he must possess adequate interrogation software to give the users real-time access to the information store in his files by means of a terminal connected to the telephone network.

This software, or command language, is a key as important to the user as the data-base itself. Two documentation collections having the same origin but exploited by different software systems will not answer the user's queries in the same way.

It should be noted that the interrogation language depends on the host and not on the data-base (with the exception of numerical data-bases and certain factual data-banks which may require a special language). This means, generally speaking, that the user of a host computer needs to learn only one command language to gain access to all the data-bases it can offer, and the host operator makes the arrangements for his training and any specialization in the use of certain data-bases.

c) Compartmentalization of the market by interrogation languages

In a competitive environment the host will naturally attempt to give himself a distinctive profile and to protect his market, if possible, by means of an interrogation language with special features designed to meet the requirements of his clientele.

In addition to the different strategies adopted by hosts, there is a more fundamental dichotomy between two types of interrogation languages which give rise to two different types of service tailored to the needs of two separate markets. On the one hand we have the elementary interrogation languages based on the structures of a decision tree and on the other we have the sophisticated interrogation languages for combinatory search procedures using keywords and Boolean logic operators such as 'and' and 'or'.

The systems based on the first type of language are intended for use by the general public, and to increase their accessibility and reduce the mistrust of informatics the computer terminal is "hidden" in an ordinary looking domestic television set. In the absence of Boolean logic and searches based on keywords, they do not permit bibliographic searches or any great precision in the selection of the factual information sought by the user. They restrict themselves to displaying summaries, progressing from the general to the particular, the user having to scan the pages to pinpoint the address of the store most likely to meet the need for information. In view of the absence of a real interrogation language, they offer access only to files of limited volume whose format has been specially designed to follow the structures of a decision tree.

The systems using sophisticated interrogation languages are of a different nature. They are aimed at professionals or at least at personnel who are sufficiently interested to undertake a training session lasting four or five days. These systems provide access to the large world-scale scientific and technical data-bases. The strength of the interrogation language permits in a few minutes the retrieval, from millions of records, of the fifty or a hundred references relating to any search profile whose subject tallies with the content of the interrogated data-base.

There are an estimated 10 000 users of such services in Europe. It should be noted, however, that these are mostly intermediaries specializing within their own organizations in the use of these systems to meet their colleagues' information requirements. This market, which will always be more restricted in size than the market for systems aimed at the general public, is nevertheless growing rapidly (few firms, at present, have staff trained to use these resources). Euronet DIANE is intended to serve it.

Euronet DIANE, the direct information access network for europe

Euronet DIANE is a European Community creation. The Resolution of the Council of Ministers of 24 June 1971 encouraging the progressive creation of a European information and documentation network originated this system, which has been developed under two Community Plans of Action in the field of scientific and technical information.

Its aim is to provide access to the documentary and factual information resources stored in the memories of the European documentation centres, by making these hosts accessible on-line to all users in the Community under equal, fair and economic conditions.

DIANE (Direct Information Access Network for Europe) is the collective name given to the information services. Euronet is the name of the specialized telecommunications network to which they are being progressively linked.

a) Why Euronet DIANE?

It is estimated that more than two million scientific communications or articles are published each year in the industrialized world. The volume of this information is currently such that the computer has become an indispensable aid to the research worker and the engineer for finding and selecting the data he needs for his work.

In certain cases, the use of sophisticated interrogation languages permits a tenfold increase, at least, in the efficiency of documentalists' and information searchers' work by reducing to a mere 15 minutes on a terminal the work that used to take half a day of manual search in abstracts journals.

However, notwithstanding the rapid development of this technique in certain countries, Europe still has a great deal of leeway to make up. In terms of demand, we are five years behind the USA and, it should be added, half this European demand for information is not met by European, but by American, hosts. The reasons for this situation are complex. Two of them are particularly significant:

- On the one hand, the absence of a specialized telecommuncations network in Europe made it *technically* difficult and economically expensive to interrogate distant hosts in the territory of the Community. It was sometimes easier and cheaper to interrogate American hosts linked to private international networks.
- On the other hand, in spite of the advantage of access to American hosts, the fact that transatlantic telecommunications cost more than cross-continental

telecommunications in the United States put a brake on the expansion of European as compared with American demand.

The continuing immaturity of the European market is also partly explained by a variety of other psychological and economic factors, but these are of secondary importance compared with the problem of access to services. Hence the concept of Euronet DIANE was developed:

- firstly, a specialized telecommunications network to eliminate the technical obstacles to access to European hosts, with reliable, rapid and efficient connexions, low charges and full compatibility with the widest possible range of equipment;
- secondly, a voluntary network of hosts who have agreed to make their services available to all users in the Community without discrimination on the basis of nationality;
- finally, the steady development of supporting activities and the step-by-step creation of common tools to facilitate the use of the network and enhance its quality.

b) Euronet, the telecommunications network

This network is specially designed for the on-line interrogation of remote computers installed and operated by a consortium of the nine national PTTs under a contract concluded with the Commission of the European Communities.

The packet switching technology employed by the network enables the users to share its common resources at the lowest possible cost while at the same time enjoying the advantages of reliable and rapid transmission.

An access point has been provided in each member country of the Community. The network will soon be extended to include Switzerland.

The tariffs are very generous and tailored to the on-line information searchers' user profile. The charges are independent of distance and vary with the mode of access, but are usually based on the length of time of the connexion and the amount of information transmitted. They encourage time spent on interactive dialogue and ensure that a fair payment is made for the volume of data transmitted.

The cost of Euronet communications for an average search profile of 15 minutes involving the transmission of 200 x 64 character segments is in the region of one US dollar.

To this 'Euronet charge' must be added the cost of the local telephone call from the searcher's locality to the national access point. For national calls to the Euronet access point, several countries, namely Belgium, the Federal Republic of Germany, France, Ireland and Luxembourg, have decided to apply a special tariff which is also independent of distance, i.e. the town from which the call is made.

In every case, the cost of telecommunications being identical irrespective of the host called, the user enjoys a wider choice, which is not distorted by the distance from the caller to the centre called. Even if the local telecommunications charge is added, the advantage is considerable in the case of an international telephone connection, as the amount saved will vary from 50-80%, depending on the country and the user profiles.

It can be said that Euronet has lifted the telecommunications cost barrier to the development of on-line documentary search in Europe.

c) DIANE - the host network

As from November 1979, six host centres were linked to Euronet, but upwards of 23 hosts have announced their intention of joining the network. They will offer a total of about 175 data-bases and data-banks in a wide range of scientific, technological, social and economic subjects.

These hosts may be public or private documentation centres, either commercial or non-profit-making. They were originally recommended by the national delegations to the Committee for Information and Documentation in Science and Technology which advises the Commission on these matters, but other candidates may present themselves directly.

The Commission signs a memorandum of agreement with each of these hosts, fixing the conditions for their connexion and the rules regarding competition, cooperation and non-discrimination. It stipulates in particular that the hosts shall bear the sole responsibility for marketing their on-line interrogation services.

These services, which compete with each other, cover such a broad spectrum that they are very difficult to summarize adequately.

Although, at first, the coverage provided by DIANE will be particularly strong in chemistry, medicine, biology, patents and science and technology as a whole, it is constantly being developed as a result of the new agreements between hosts and database producers (the majority of the new files announced in the past three months relate to the economic and financial sectors which are steadily increasing their share in the range of services offered by the DIANE hosts). There is still much ground to make up in the areas of cultural and artistic information, which is currently represented only by the major national bibliographies based on the information provided in the various countries by the national book depositories or similar sources.

The most frequently mentioned data-bases are those of American origin. However, 70% of the files announced on Euronet DIANE are of European origin. As most of these are new, their scope is naturally more modest than that of the American files and they cannot all be accessed on-line at present.

It is difficult to give a precise idea of the costs of interrogating these services as they vary considerably between hosts and between databases and, furthermore, according to the type of subscription the user opts for. The tariffs themselves may be different, particularly for access to numerical data-banks. In the case of bibliographic files, the majority of hosts apply a tariff system based on the time of connexion and the number of references printed out off-line. To this charge, which includes the remuneration of the data-base producer, must be added the telephone charges. In the case of the DIANE hosts, the charges are expected to be somewhere between $50 and $100 per hour for the majority of bases.

It should be noted, however, that this charge can vary by up to 100%, on the same host computer, depending on the data-bases and the charges made by the producers and that, generally speaking, prices are going up, with the majority of new files covering the economic and financial sector being offered at an average price of about $100 per hour.

DIANE does not represent the merger of the European documentation centres linked with Euronet, but rather their permanent association around a specialized telecommunications network which respects their individuality and personality.

It is this concept which gives DIANE its momentum and richness; but it also introduces an element of disorder which could be harmful to the growth of the market because the potential user may be disconcerted by the multiplicity of services, languages and interlocutors or by the very fullness of the documentation bases that are available.

That is why the Commission, in cooperation with the hosts, has launched various supporting activities.

d) Supporting activities

Apart from technological support, i.e. action to strengthen the network itself, the Commission is taking an active interest in the development of the scientific and technical information market in Europe.

It therefore endeavours to promote a dialogue with the users in order to identify and meet their requirements. To this end, a Users' Forum has been established and will meet periodically in Luxembourg.

Apart from its dialogue with the present users of on-line information systems the Commission also makes every effort, of course, to inform potential users of the availability of these means of interrogation and of their advantages. With this in mind, it has launched a promotion campaign for DIANE and Euronet services. This has so far been exclusively aimed at the professional operators in the information and documentation sector.

It also intends to help the network's users by facilitating the conditions of access and use. Hence the creation of a common interrogation language to permit the interrogation, with the same basic commands, of bibliographic files divided among different host computers. This optional language will not be provided on all the hosts, but it is already being incorporated in most of the software available on Euronet DIANE. Studies of the possibility of developing a similar common language for access to factual and numerical data-banks are currently underway.

In a similar vein, an automated information service provides on-line information about the network and the available data-bases. This entirely free service (apart from the cost of the local telephone call to the Euronet access point) is accessible in the six Community languages. A more ambitious service, a veritable data-bank of the data-bases available on Euronet DIANE, is envisaged for a later stage.

This does not mean that every obstacle to the development of the market has now been surmounted; other developments are still necessary.

There are, of course, the linguistic barriers; multilingual thesauri and files can be an aid to removing them, particularly with the assistance of the machine translation systems which are now becoming available.

There is also the problem of the commercial fragmentation of the hosts' service, which obliges users to conclude a contract with every host to which it desires access; although there is a common command language on DIANE, there is not yet a single contract and single password for access to the various services. Progress in this case will be achieved step by step.

Downstream of the hosts, the brokerage market should be encouraged, to enable small and medium-sized firms and other undertakings to profit from the development of the information resources available on-line without having to acquire a terminal or train one of their staff to become a specialist.

Upstream, there are new files to find and render accessible, particularly in the form of factual data-banks in the social and economic sectors, to meet the rapidly increasing demand in this area.

Euronet DIANE and the outlook for traditional operators in the information market

Although Euronet DIANE is in itself a modest project of minor dimensions, its impact has been felt far outside its specific sphere of activity. As an international network, it will transmit only a small part of the on-line interrogation traffic, the part corresponding to the imports and exports of the ten countries linked with the network. But it has transcended that simple function and has served as the catalyst for highly important technological and organizational developments in the realm of information. In the member countries, the introduction of Euronet stimulated the creation of a number of host centres such as BLAISE, INFOLINE, TELESYSTEMS and SPIDEL. Furthermore, the opening of Euronet ensured public access to a number of other services hitherto monopolized by a closed group of users. The common tools, e.g. the common command language and the multilingual thesauri, that have been forged to facilitate the use of the network have had an effect on the whole information market which goes far beyond the Euronet traffic. Likewise, the Commission's promotion campaign has stimulated not only the exchange of information within the Community but also the development of the national information markets.

a) The risks

For the publishers of books or scientific journals, the main risk is that of losing a unique opportunity of benefiting from the development of new techniques of dissemination. It is highly improbable that the expansion of on-line interrogation will lead to the decline of printed publications. Indeed, the increased use of bibliographic files will boost the demand for primary documents. The factual data-bank, for which the charge is $100 per hour of interrogation, cannot really be regarded as a competitor of the printed book, because it meets the needs of a category of users who could not possibly manage to gather together, in printed form, the entire documentation base needed to cope with the variety and range of their queries.

The only real potential overlap might be in the field of abstracts journals or indexes, but the printed bulletin will undoubtedly continue to coexist with on-line documentary search systems, perhaps at the price of certain changes in its role, as it will provide a low-cost response to repeated and specialized information needs.

b) The prospects

On-line bibliographic searches generate a demand for primary documents which would not otherwise exist. Various operators have a stake in this and their activities may conflict. A study of document delivery, carried out by the Commission, and the discussions it has provoked, have demonstrated the particularly delicate nature of the problems raised by this question, particularly that of copyright. Whatever solutions are considered, the traditional operators must equip themselves to meet the demand for primary documents engendered by the interrogation of bibliographic files. The control or limitation of illicit copying is of no real use if the conditions of access to primary documents do not allow this demand to be met. This phenomenon may possibly necessitate the modification of certain methods of production and publication and it may be better for publishers, in certain cases, to hand over to other operators, in accordance with appropriate arrangements, the task of meeting the demand for retrospective documents.

Another field of activity now open to publishers is the creation of new data-bases and data-banks, where they can draw on their experience of collecting, sifting and processing primary information or, quite simply, where they can enable the host operators to exploit the traditional documentary resources that have already been stored on magnetic tapes but have not yet been edited for direct on-line access.

Euronet DIANE and the similar systems using sophisticated languages do not, therefore, constitute a threat to the traditional operators in the information market. Indeed Euronet DIANE could well serve as a source of growth and prove the ideal medium for risk-free familiarization with telematics and the new technologies providing access to information.

Origination and Editing: the point of view of the author, editor and referee

Chairman: Klaus SAUR
Rapporteur: Edward PHILLIPS

In future authors will have the opportunity of being much more actively involved in the dissemination of their work. This development, a result of the new methods of information handling, storage and transmission, was seen by the participants to be at the root of the changes which are beginning to occur in the editorial side of primary information publishing. Already it is commonplace for authors of journal articles, and of books, to be asked by their publishers to submit final manuscripts in the form of "camera-ready copy", by which it is meant that the typescript should be well enough laid out and sufficiently free of typing errors for it to serve as a "master" from which the printing plates can be formed by a photographic process. By making authors thus responsible for the printed text, the expense of having their typescripts set in print and the drudgery of proof-reading are eliminated.

Such tendencies will intensify and will have important repercussions in the printing industry. Already there is resistance to author-prepared "camera-ready copy" from the traditional typesetters, whose role in the printing process is being bypassed. In this first session, however, the discussion concerned the implications for the editors, the referees who are called upon by editors to aid in the assessment of the merits of material offered for publication, and the authors themselves. In other words, it was limited to the effects of the new technologies on the group of people who decide what material is to be disseminated.

It was clear from the talk by Mr. Page that the potential already exists for major changes in the working habits of authors, editors and referees. All stem from the possibility of having the author's master text transmitted from place to place by electronic means, over the telecommunications network. At each location, whether it be the author's, the editor's or the referee's office, the text can be displayed on a screen and modifications can be made to it through word processing equipment. At no stage need the text be typed on paper, and of course any modifications which have to be made to individual words, lines or paragraphs can be done without retyping the whole text. Finally, when the text for publication is agreed upon, the stored material can be typed out to produce the "camera ready" master or serve as input for photocomposition, or it can be switched directly into an information network for "electronic" publication, so that users of the material will receive it on a screen, rather than in printed form.

That such possibilities will become realities was not doubted by the participants. The individual elements of the system already exist: it is not hard to imagine that the word-processor terminal will become almost as ubiquitous as the electric typewriter, and that information networks will be able to handle vastly more data than at present. But efficient as the new methods might be, some participants saw that certain changes might result which are not altogether desirable, implying that the traditional methods for the dissemination of information have some advantages which in future may be lost.

Drawbacks of the new technologies

These reservations generally had to do with the quality of the published material. Already there is a tendency for editors and referees to keep to a minimum the changes which they request of an author who has submitted a paper in the form of a camera-ready typescript for publication in a journal which uses this method. Obviously it is much more time-consuming to alter such a typescript than it is to alter one which will be sent to a typesetter for setting in print in the conventional way. This trend towards the avoidance of modifications could well intensify as editors and referees are increasingly faced with texts submitted in an electronic form such that they could be sent for publication at the touch of a button. The temptation to approve for publication texts which deserved drastic rewriting may be too strong to resist, some participants feared, resulting in a deterioration in the standard of writing. Already many users deplore the lack of clarity and brevity which they claim mars much of the scientific and technical literature, for example, so there was a sentiment among some participants that fears of further deterioration should not be taken lightly. A wider-felt apprehension was that the scientific and technical quality itself might also suffer for the same reasons. Indeed, the impact which the new technologies might have on the quality of the content of the published material, whether it be scientific, technical, economic, social, or information from some other domain, was a major theme running through the discussions.

Fears also centred on the possible consequences for the information publication industry as a whole. The new technology was seen as being costly and there would be a tendency towards concentration; the position of the big publishing houses would therefore be strengthened, at the expense of the small publishers. Such a trend would be deprecated by many of the participants, who considered that the small publishers add colour to the publishing scene, can be a source of new ideas, and are a safeguard against the covergence into a few hands of control over what material is to be disseminated. It would be a serious loss, it was felt, if small publishers were "rationalized" away, a sentiment which applied to the publishing industry as a whole, not just the information publishers.

It was also asked whether for reasons of cost the new technologies would be limited to the rich countries, adding to the isolation of poorer countries in, for example, scientific and technical matters. And even in the rich countries, the cost of the equipment might be such as to limit it to the richest organizations and institutions, thus creating a "charmed circle" having access to the information circulating in the system, which is denied to the rest of the population.

Authors and referees

Apart from fears about the extra pressures which referees might be subject to, the participants had little to say about the impact of the new technologies on the referee's role. Although in future he might judge a submission which is displayed on a screen rather than arriving as a typescript on paper, the essential task of the referee will remain that of assessing the intellectual quality, the clarity and the relevance of the material offered for publication.

Looking at the new technologies from the point of view of the author, however, the participants saw more likelihood of significant changes in methods of work. Opinion seemed to be divided about the acceptability of these changes. One view was that authors will find it difficult to compose their works on a word processor, and will continue to prefer the "pencil and paper" method of composition. It is widely believed that when it comes to the drafting of a text for publication, many authors prefer to retreat from their place of work to the peace and quiet of their own homes, which at least in the short term are hardly likely to be equipped with word processors. Indeed for these reasons one participant made a plea for the retention of traditional methods, which he termed the "Gutenberg" system of publication. It was further suggested that the supply of information to authors about the new systems, their methods of use and potentialities, is lagging far behind the development of these systems. As a result authors are already bewildered by the options available to them.

In essence this body of opinion seems to be saying that though the introduction of new technologies at the author's end of the publishing process is inevitable, authors will not give their immediate whole-hearted acceptance. Authors would however find it easier to cope, it was pointed out, if more and better information could be made available to them. This view was shared by those participants who took a more positive view of the acceptability to authors of the new technologies. One participant asked that more space be found in higher education curricula for teaching on modern information science in general. Another felt that the new technologies based on electronics and telecommunications are going to play such an important part in all aspects of living that teaching about their use should begin in primary schools.

The question was also raised of the "prestige" of the new methods of publishing. The method of assessing the worth and merit of an academic by totting up the number and extent of his publications is widely frowned upon, but is also widely believed to be common practice, especially as an aid to those called upon to assess candidates for promotion. Will an "electronic" publication be considered as second-rate in comparison with a conventional publication, or will the two types of publication be regarded as carrying an equal status? Or, as one participant suggested, will the new systems become merely a sophisticated method of circulating preprints, which eventually will find their way into conventionally printed journals?

Another group of participants preferred to lay emphasis on the advantages which the new technologies offered to authors. Some considered that greater control over the dissemination of information would accrue to the authors of the information than is at present the case, and herein was the chief attraction as far as the authors were concerned. In other words, the publishers might find their role tending to be bypassed. It was foreseen, for example, that having composed a text to his satisfaction and stored it in the memory of his word processor, the author might then, simply by pressing a button, be able to make the text generally available through the telecommunications network to anyone with a suitable terminal. As Mr. Stephens explained in his talk, such a facility is already used by the "information providers" of the Prestel system.

Thus the author can publish without the need for an intermediary, he can himself set the fee which readers must pay to consult his work, and, as Mr. Stephens mentioned, distribution through the telecommunications network allows the possibility of counting the number of times that the work is consulted. A closer relationship between authors and readers is likely to be a major benefit.

There may however be a price to be paid for the freedom thus conferred on authors. Many participants foresaw that the opportunity for disseminating material without the constraints imposed by the referee system would be hard to resist for many authors, and the result could be a swamping of the electronic means of publication by information of poor quality. The sifting of this material can only be done at a cost which will have to be borne by the user. One participant went further, in identifying a threat to society in general from indiscriminate dissemination of material without any control on quality: he feared that persons in positions of power may be led to apply wrong policies on the basis of incorrect information circulating in the system.

Once again, then, the question arose of quality control. Several participants pointed out that the problem already exists, and has done for years. The users of scientific and technical information, for example, notably the libraries, learn by experience where valuable information is to be found. Publishers of journals which accept papers of poor quality will find that subscriptions are difficult to sell. Market forces, then, ought to work in favour of material of good quality.

Publishers and editors

In future publishing systems, will market forces continue to favour the good at the expense of the mediocre and downright bad? No clear-cut answer was arrived at during the session. But the existence of such fears about lack of quality control encouraged many participants to say that publishing houses, instead of withering away as the new technologies develop, will continue to have a central part in the dissemination of information. As now, the publishers will manage a system of editors and referees whose job will essentially be to select and "package" what is to be disseminated from the information submitted to them. (And if publishers continue to exist they will no doubt continue to commission works on subjects which have been neglected in the literature but for which they nevertheless believe a market exists.) What will change is the method of dissemination, from the present "print on paper" to electronic means. But it can be argued that this is a change of technique rather than of principle: printing has always been just one of the tasks of the publisher to be set alongside the selection and editing of material, promotion and marketing, and it is in these areas that the publisher's expertise will, according to some participants, continue to be in demand.

Within publishing houses, assuming that in future there will be no serious by-passing of their role as a result of direct publishing by authors themselves, the task of the editor, like that of the referee, is expected to remain unchanged in principle. In practice however it is possible that the interaction between author and editor will speed up because of the new telecommunications; both parties will find it easier to clear up misconceptions which arise in the interpretation of a text, for example. It was also predicted that new information retrieval techniques will prove of value to the editor, notably for finding new authors and referees. Indeed, one view was that by generally making the interaction between authors, editors and referees work more efficiently, the new technologies will tend to improve the control of quality rather than sabotage it.

To sum up, the main threads running through this discussion session were:

- Quality control: the new technologies may be conducive to a lowering in standards, although several participants considered such fears as unrealistic, and the view was even expressed that there might be a tendency towards the raising of standards.
- Changing role of the author: the author will probably become more closely involved with the publishing process, either through a publishing house or by being able to enter his work directly into the system without going through an intermediary.
- Concentration: the cost of the new systems may be high, thus encouraging the formation of groupings of publishers and hence the concentration into fewer hands of control over information dissemination. The circle of users may also be limited for cost reasons, thus cutting off poorer users from access to the system.

Processing: the point of view of the publisher, printer etc.

Chairman: D.H. BARLOW
Rapporteur: Edward PHILLIPS

"Like opening Pandora's box" was the chairman's description of his reactions to the issues being aired at the symposium. Among the beckoning opportunities arising from the new technologies, there are also many challenges and difficulties to be faced. To begin with, the chairman recapitulated the characteristics of what he termed the "ultimate" publishing system, which is being made possible by developments in computers, word processors and telecommunications. Authors will have direct access to the system and the material which they enter will be transmitted electronically to the editor, who can arrange for refereeing to be done on-line. When all concerned have had their say, the final text can be switched to drive a film-setter and hence generate the published product.

Market pull or technology push?

Did the participants agree, the chairman asked, that present trends are indeed leading to such a system, and if so, what are the reasons behind the trend? Are the users (the authors at one end of the process of information dissemination and the readers at the other) pushing the publishers, or is it that the economics of the new methods will be so much more favourable than conventional publishing methods? Or again, could it be that there is a tendency in mankind always to use the most advanced technology that is available to accomplish any given task - without giving too much thought to its appropriateness?

At this point an important distinction can be made between two ways of applying the new technologies. Its use could be limited to what is termed the "manufacturing process" of publishing, in which case the effects on the authors and readers need not be particularly marked. The publishing process in this case could be treated as a "black box", the author entering information which is processed within the box and eventually delivered to the end user, neither the author nor the end user being concerned with or perhaps even aware of the mechanisms inside the box. In other situations the new technologies are giving rise to entirely new information systems which will affect the users and the originators of the information, and will involve them in new methods of working.

In answer to the chairman's question, it is the economics of the new technologies which will attract the publishing industry, according to many of the participants. From the point of view of the physical "manufacture" of the published product, publishers are being forced to adopt new methods on the grounds of cost, said one speaker, who went on to remark that he nevertheless welcomed the change because he found that the new methods made his work easier. This view was echoed by another speaker who disposed of the notion that changes are being forced on an unwilling publishing industry: there are no developments in use, he claimed, that the industry did not want or was not willing to pay for.

From the users' position, it can be argued that the authors are reasonably well-served by conventional publishing. They see their work produced in a satisfactorily attractive form, and brought to the attention of the desired audience by the traditional methods of distribution. Perhaps the only extra which authors would ask for would be more rapid publication than is the norm at present. The readers are probably satisfied too: the much valued browsing facility is available through the supply of learned journals, retrospective searches can be done using the secondary services, and documents are easily obtained in the form of photocopies. Any pressure for changes therefore will come from the publishers rather than the users.

It was also pointed out that the originators of the information considered during this meeting are generally not professional authors. The papers, articles and books which they write are by-products of their main activity, which is that of research in their chosen field. It would be wrong to force these authors, who probably write very few works in their lifetime, to come to grips with the new technologies. It should be sufficient that the author offers a clear manuscript to the publisher, who disseminates it using the most appropriate technology.

Another participant put forward what seemed to be a diametrically opposite opinion. The real advantages of word processors, it was suggested, are only to be won if they are installed where the authors are, and if they are used by the authors to generate the material offered to the publisher, thus alleviating the publisher's production costs. Otherwise it is the publisher who is responsible for text composition, just as in "traditional" publishing - he may use new techniques, but his task remains fundamentally the same as it always was, and the new techniques do not confer any really major benefits.

On balance the opinion was that it is likely to be the economic aspects of the new technologies which will make them attractive to the publishing industry. Pressure from users will probably play a much smaller part.

Fate of the specialist journal

One factor pushing the publisher towards the new technologies because of the economics which might thereby be achieved comes from the libraries, which are of course the main buyers of learned works, particularly the journals for users to browse through. Journals for browsing are however the first casualties when pressure is applied to library budgets, and it is the specialist journals which will suffer particularly badly. Many might well be forced to cease publication, unless they can be produced by an alternative to conventional methods, such that their cost to the library is held down and thus subscriptions to them are less open to cancellation. Electronic publication might then turn out to be a way of saving the specialist journal, which is in any case probably more suited to this publishing technique where future telecommunication developments offer the possibility of linking authors, readers and referees who have similar interests. On the other hand the larger, more general, journals will continue to survive for some time in their conventional form, it was suggested.

All will not be plain sailing however for publishers who embrace new technologies. Labour problems were cited, one participant remarking that publishers have already been held back from a deeper involvement in the new technologies for this reason. There seemed to be some doubt, however, as to whether the new technologies will lead to significant redundancies in publishing offices, or whether it will be just the nature of the various publishing jobs which will change without any great reduction in their number. The fears are real, nevertheless, and account has to be taken of them.

It was also suggested that a heavy dependence on computers and telecommunications will make the publisher more vulnerable, to industrial action in the telecommunications industry for example. The greater the publisher's outlay on new technologies, the more strongly he is locked into the system. For this reason one participant foresaw publishing developing in three main directions. There will be large commercial publishers who will rely heavily on the new technologies and pay the price in terms of vulnerability. Second, some small publishers will be able to make a living by concentrating on rapid publishing using traditional methods. Third, harking back to the fears mentioned in Session II about the dangers of information of poor quality, this participant thought there would be publishers who would use the new technologies to increase the proliferation of unrefereed material.

Photocopying and the role of the publisher

One technological development in particular - photocopying - has of course been a worry in some publishing circles for years, and was not neglected during this discussion session. One participant who expressed himself forcibly on this subject argued that it was not because of their production costs that the specialist journals, and their publishers, will go out of business. Rather it is photocopying, coupled with the tendency for libraries to cancel journal subscriptions in the knowledge that requests for the cancelled journals can be met by borrowing, or by obtaining photocopies, from one of the big centralized libraries. Publishers are thus being undercut, as it were, by the big lending libraries. It is not only the academic libraries which are following this practice - the libraries of industrial enterprises too are increasingly relying on photocopying to the detriment of the journal publishers, a fact which the speaker thought was particularly reprehensible when, he said, much of what is referred to as industrial research is in fact the development of new products. Publishers have no defence, he went on, when advantage is taken of them in this way.

Another speaker, agreeing with this view, thought that the situation could be turned to the benefit of the learned societies. Rather than use the specialist journals of commercial publishers, authors might be persuaded to deposit their work in the archives of the relevant society or institute, which would make a charge for the supply of copies to users on demand. Such a system might be a useful source of funds for many societies which are on the verge of extinction.

This notion was attacked on the grounds that learned societies might refuse material which in some way did not conform with the opinions of the societies themselves - a danger of censorship in other words. Such dangers would arise, it was said, whenever corporate bodies, through their ability to afford the necessary capital expenditure, achieve control over the flow of information in a particular subject area. Organisations such as the major electronics corporations, the broadcasting and the telecommunications authorities, as well as the learned societies, were cited as examples of bodies which could conceivably monopolize the dissemination of certain types of information. Competition is therefore essential in publishing, so that new ideas and information can reach the public without hindrance.

This then was an argument in favour of the publishing houses embracing the new technologies, in spite of the many difficulties and costs involved. The maintenance of the publisher's role would in fact be in the interests of the users of information, it was suggested. Given the vast growth in the amount of information available and in our ability to store and retrieve this information, some guidance is vital to help the user reap the full benefits. Such assistance comes from the publisher who, as well as offering some guarantee of the quality of the information which is offered (through the refereeing procedure for example), has the experience to disseminate information in the manner which is best suited to the needs of the readers and the authors. It would be wrong to pretend that these services can be provided cheaply. The new technologies will demand major capital investments on the part of the publishers, yet at the same time these technologies make it easier for information to be copied without the authority of the author and publisher of that information.

But, the same speaker continued, if the publisher shuns the new technologies, then others will step in, such as the aforementioned broadcasting and telecommunications authorities. It is important that the publishers maintain their position in the information industry, and that there should continue to be a place for the small publishers for the reasons mentioned in Session II.

Indeed, another participant pointed out that within some large corporations there are already a variety of well developed and efficient systems for the transfer of information. The publishers might well learn from a study of these systems: it is conceivable however that such systems could also develop so as to enter into competition with services provided through the publishing industry.

New products

What kind of products will the publishers be offering in the future, as a result of the new technologies? In answer to this question from the chairman, it was pointed out that any new product must either be a better proposition economically than the traditional products being replaced, or it must offer some new dimension which cannot be provided using traditional methods. New technologies make it possible to retrieve information in a way that is impossible using conventional methods, for example. But the participants found it difficult to be precise about what types of new products one might expect to see in the future. As one speaker pointed out, it is difficult to make cost comparisons between present information products and their possible future replacements, if only because we cannot be sure about the costs involved in the new systems based on electronics and telecommunications. And of course acceptability to the user is often difficult to assess in advance: microform, for example, is a cheap and apparently convenient method of carrying information, but, as everyone knows, it has met with a great deal of resistance on the part of users. Another example cited is the computerized data-bank, which generally is also available in conventional printed form. Hardly anyone in the information industry, said the same speaker, is making a profit out of a product such as a data-bank which is exploited only electronically. The problem for the enlightened publisher, he went on, is how to adopt a stance enabling him to exploit the new technologies when the future becomes clearer.

At this stage participants were reminded that new products designed to take advantage of facilities in the technologically advanced countries may not be suitable in the developing world. The obstacles to the dissemination of medical information were cited. It is difficult enough to get medical information in the form of books, in the right language, into the hands of the crucial middle and lower echelons of the health administrations in many developing countries: it will be a long time before the new media become the appropriate carriers for information aimed at these destinations.

Need for variety

A related comment was that there is room for a range of methods of disseminating information. For "pure data", an electronic system may well be the most suitable. But one can imagine types of information which might be better suited to dissemination on microform, and in some cases paper will be the best medium (for information concerning the social sciences and the arts, it was suggested). This remark echoed another comment, which was to the effect that the introduction of the new technologies, making use of the most modern developments, will not necessarily result in the disappearance of the more orthodox printing and publishing techniques. Much will depend on the costs of the new methods as compared with conventional techniques, and it will no doubt turn out that various media will be in use simultaneously, ranging from the most advanced "viewdata-type" system at one end of the spectrum, through the various new printing methods, to traditional printing methods at the other, each doing the job for which it is most appropriate.

The main ideas presented during this session can be summed up as follows:

- Motive for change: the economic advantages of the new technologies, as perceived by information publishers, will probably play a greater part in promoting change than pressure from authors or users.
- Copying: major capital investments on the part of the publishers are demanded

by the new technologies yet these same technologies, by encouraging copying, make it harder for the publisher to protect his investment.

- Role of the commercial publishing house: it is important for society that no monopolies exist in the information industry, and this can best be guaranteed by ensuring that publishing houses can continue to flourish.
- Dominant technologies: no single medium or technology will achieve a dominant position. There will be a variety of methods in use, to suit different types of information and the needs of different types of users, and one can expect to see highly innovative techniques existing in parallel with conventional printing and publishing methods.

Dissemination, part 1: the point of view of the secondary services

Chairman: C. SALMON
Rapporteurs: J. Michel GIBB and Marcel MAURICE

The chairman suggested what meanings should be given to the item "secondary services" for the purpose of the discussion. He settled for a broad definition, covering:

1. The private or public enterprise institutions which identify primary documents (i.e. documents containing original information relevant to scientific, technical, economic, social problems, etc.) within a predetermined subject field, abstract them, index them, thus generating bibliographic data-bases from which various products or services can be derived and marketed, e.g. a traditional abstracts journal on paper, a magnetic tape service, an on-line service (usually via the large specialized intermediaries) etc. thus permitting users to find out about documents which they should read or consult.

2. Large documentation centres, such as those operated by professional associations, which generate their own data-bases, tailored to the highly specialized requirements of their members, and which, with the aid of their in-house computer facilities, conduct retrospective searches or current awareness services on the basis of individual customer "profiles".

Five main themes were proposed for discussion.

With the gradual introduction of "electronic journals" (i.e. journals which are made available on-line via a screen or a teleprinter, etc.), will the respective roles of primary publishers and secondary services be seriously modified?

The speaker who introduced this theme assumed that publishers wishing to try out the electronic journal concept would issue synopses rather than full-length articles, presumably because this would be a cheaper approach initially. This led, later in the session, to some argument as to what a synopsis actually is, as opposed to an abstract. Although not, at first sight, strictly relevant to the main point, this part of the discussion proved rather illuminating and will be reported below.

First, let us turn to the main theme. A further assumption was made by the main speaker: instead of being presented in packages within the framework of a journal, the electronic synopses (or full-length documents, for that matter) would tend to be offered by publishers in disarray. Potential customers would however register their "profile" with the publishers and would only need to consult articles fitting in with this profile.

This however does not seem very practical (imagine having to register your profile with several dozen publishers!). A new brokerage function would therefore have to be fulfilled, that of intermediary between the primary publishers on the one hand and the secondary services on the other. His job would be to look through the monthly mass of primary material on offer and channel the "right" articles to the "right" secondary services, which would then in turn play their traditional role of helping users identify documents of interest to them.

This possible scenario for the future was not further discussed, perhaps because it did not appeal to the audience. Its author himself pointed out one of the dangers which it would involve, namely the potential degradation in the status of electronic articles, which could gradually come to be regarded as "grey" or non-conventional literature.* In the opinion of the rapporteurs, this very fact requires that the proposed scenario be submitted to further careful scrutiny.

The problem would appear to be to obtain the ease and speed of access opened up by the use of computerized systems, allied to telecommunications networks, without sacrificing the quality control exercised within the environment of the traditional journal, thanks to the interaction between authors, editors and referees. It is thus important, even if a certain amount of grey literature is to be published electronically and announced by secondary services, to be able to distinguish it easily from the high quality literature today published in "core" journals.

This being said, several speakers insisted on the importance of the grey literature, which often contains valuable information (if one is prepared to wade through a mass of material in order to find it), but remains largely unknown because it is not distributed through usual commercial channels. The secondary services should rise to this challenge by extending their coverage of this type of literature and thus helping the public to secure better access to it. It was recommended that a comprehensive solution to this problem should be sought at a European level. In fact this is already happening.**

The answer to the initial question thus seems, on present evidence, to be that no major changes will occur in the respective roles of primary producers of literature and secondary services. However more collaboration between primary publishers and secondary services was urged, if only for the sake of speed of announcement. Already today, some publishers have entered into arrangements whereby they provide secondary services with abstracts and bibliographic data as soon as a document reaches the proof stage. In an age where more and more literature is generated in machine readable form and where telecommunications are becoming easier, it should become easier and easier to do this. Some publishers may be afraid to do so, of course. These fears were analyzed later on.

* *The "grey" literature is that which is not published through commercial channels, including laboratory reports, a large number of official documents, theses, etc. and therefore is unlikely to have been through the editing and refereeing procedures to which conventional literature is usually submitted.*

** *See Gibb, J.M. and Phillips E., "A better fate for the grey, or non-conventional, literature", Journal of Research Communication Studies, I, 225-234(1979).*

What new problems will the secondary services encounter in a context of "electronic" integration?

If it is true that the whole process of generating primary information, starting with the author himself, right up to its consultation by the user, is to be integrated thanks to the "electronic" revolution, what special consequences is this going to involve for the secondary services?

At first sight, this new context promises to be beneficial to them. The traditional way of producing bibliographic data-bases involves not only long delays but duplication of work between the various stages in the information chain: for instance, the title and abstract, already keyboarded at the primary publication stage, have to be entered completely afresh by the secondary service, or the several secondary services (each working independently, of course) interested in announcing the document.

The speaker who introduced the theme expressed doubts about the practicability of integration, not only because of purely technical compatibility and reliability problems, but because of the sheer complexity of the organizational patterns that would have to be adopted.

He further expressed the fear that primary publishers might be tempted, out of prestige reasons, to generate their own small data-bases in order to advertize their own products. Moreover, some secondary services, in order to reduce technical problems to a minimum, might be tempted to restrict the coverage of their data-bases to the output of the larger, well-equipped primary publishers. No one in the audience seemed to share these fears. Small data-bases or those which do not offer the guarantee of covering a high proportion of the available literature relevant to their declared subject scope are not likely to attract many customers. Thus sheer economic factors will hamper their emergence.

Nevertheless, one speaker pointed out that, even today, there is a proliferation of bibliographic services in certain fields (he was alluding to the social sciences, in particular), with the result that the same document is often abstracted several times over. This, in itself, does not seriously inconvenience the user. The trouble is that the bibliographic records and abstracts of one and the same document differ so much from one another that they tend to confuse the user completely.

Another speaker retorted that there was nothing basically abnormal or undesirable about this. Depending on the point of view of secondary services (and therefore of its customers), abstracts can very well present different pictures of one and the same document.

What to conclude from these apparently contradictory positions? Perhaps one provisional conclusion can be drawn, namely that it is not only impractical, because incredibly high standards of discipline would be required, to arrive at a perfectly integrated system, but it is probably not even desirable in the first place, in a society which prides itself on its ability to cater for individual taste.

Will the demand for paper products (abstracts journals) of the secondary services persist, or will users prefer the electronic method of consultation?

No one was prepared to commit himself to a clear-cut answer to this question.

An allusion was made to the International Food Information System, as an example of a secondary service which provides a full range of alternative products: 1) an abstracts journal; 2) a tape service for the benefit of organizations wishing to conduct searches themselves; 3) an on-line retrieval service. All that can be said at the moment is that these different services advertize each other, with the result that the demand for all of them tends to be growing. It is however probable that with the advent of better telecommunications facilities, via Euronet, for instance, the competition of the on-line service against the other products will become sharper. Moreover, the electronic medium may prove the most attractive for services which provide short-lived, e.g. current awareness, information.

A speaker experienced in the publication of bibliographic reference books issued a word of warning, taking as an example the case of "British Books in Print". This work, invaluable to booksellers, used to be published in book form only at yearly intervals. A microfiche edition, updated monthly, is now also available. It is interesting to note that users prefer to use the book (you can thumb through it easily, you can move it about from desk to desk, etc.) rather than the microfiche (although it provides more up-to-date information). Sales of the book have therefore been unaffected and remain at well over 20,000 copies, whereas only 800 customers have been found for the microfiche service. The lesson to be drawn from this, according to the speaker, is that offering alternative products could become a risky commercial proposition. In the case of British Books in Print, no harm has been done but other similar ventures have led to a situation where, instead of one product being sold profitably, two alternative products are each being marketed at a loss.

In the view of the rapporteurs this does not mean that the approach of secondary services should be conservative towards the question of diversification of products, but merely cautious.

Will the role of secondary services gradually decline, until they disappear?

The speaker who introduced this theme alluded first, for the sake of example, to the fears currently expressed by some documentalists (information officers), who feel that with the prospect of increasing availability of on-line systems, both for search purposes and then for obtaining the full text of the documents required, they will in due course become redundant.

It is alleged that many users dream of being able to conduct their own on-line searches themselves.

In the opinion of the rapporteurs, only a handful of research workers are likely to feel strongly about this. Most users prefer to have their information presented "on a plate" and are prepared to pay for this.

The speaker likewise put forward reasons for dispelling these fears. Conducting a satisfactory search requires a good general knowledge of one or several related subject fields, which an information officer normally has because of his training and experience of following up different queries, otherwise important items will be missed; as the average user does not have this wide knowledge, the information officer is a useful ally. Moreover, the information officer has the opportunity of becoming familiar with the jungle of different possible sources of information, as well as with the best method of getting the maximum number of relevant references out of this or that data-base. Finally, he is expected to have a greater knowledge of foreign languages than his customers; this again increases his ability to pinpoint useful information.

Similarly, it is clear that the producers of bibliographic data-bases, if they are doing their job well, are detecting as many suitable items of literature as possible and indexing them carefully in order to make their retrieval easier. The time and expertise devoted to this is an investment which provides invaluable dividends to the customer subsequently.

Will the growth of electronic secondary services threaten the economic viability of primary publishing?

The first statement made in this connexion supported the view that, on the contrary, secondary data-bases advertize primary publications, and that this trend is on the increase. In the past quite a number of bibliographies gathered dust on library bookshelves. Today and tomorrow, thanks to the growing efficiency of on-line bibliographic systems, more people will become aware of more primary literature likely to be of interest to them. From the primary publisher's point of view this is at first sight a favourable development.

As could be expected, a number of publishers rightly objected that this was only a partial picture. Admittedly users are made aware of interesting published documents, but where do they order them? In many cases not from the publisher but from a library equipped with good photocopying services. This kind of practice blatantly violates copyright law. It was decided to postpone discussion of this problem, which is primarily one that concerns publishers and booksellers, to the following session.

It is however interesting to note that at least one secondary service active in the legal field (and thus no doubt particularly conscious of the copyright situation) in a European country has gone to the trouble of setting up a bookselling department which handles requests for documents in a normal commercial way.

One speaker, wishing to dispel a misconception which he suspected was growing in some participants' minds, raised the issue as to what a synopsis is, as opposed to an abstract (synopses were mentioned earlier in the session). Abstracts, and not synopses, are traditionally provided by secondary services. Their purpose is not to provide actual information but rather to indicate to users whether they are likely to find information relevant to their needs in the complete document.

The speaker, a librarian, claimed that many abstracts fail to achieve even this limited objective. However this may be, he pointed out that electronic communication networks like Euronet were, as things stand, destined to give access to secondary databases and thus to be carriers, in effect, of "null-information".

This criticism is not as scathing as it sounds, but it makes it useful to have another look at the synopsis concept. The speaker provided the following definition of a synopsis: 1) it should provide the names of all the authors; 2) it should present, in a few hundred words, the vital information contained in the full-length document; 3) it should give references of all documents cited in the original.

The rapporteurs have come to the conclusion that, as was pointed out by the speaker who introduced the first theme of this session, the synopsis concept may well be highly suitable for electronic publishing of primary material (a synopsis *is* a primary publication, unlike an abstract). This would not in any way interfere with the secondary services, which would continue to help users to detect relevant information, and would mean at the same time that the information content of the data carried by telecommunication networks could be enriched.

Dissemination, part 2: the point of view of the publisher, bookseller, librarian, etc.

Chairman: J.K.W. VAN LEEUWEN
Rapporteur: J. Michel GIBB

The chairman proposed that the point of view of the bookseller be examined first, followed by that of the librarian, the end-user and the publisher.

The bookseller

Will booksellers disappear in due course as intermediaries in the information process, because of the introduction of new technologies?

No one claimed that this was likely to happen. On the contrary, the general view was that they would continue to play a useful role.

It should be borne in mind that the introduction of electronic publishing, which would theoretically enable the booksellers to be by-passed, will only be gradual. In any case, electronic publishing is not likely for a long time, if ever, to affect books, which remain the major source of information for the general reader. Keeping track of the hundreds of thousands of titles published in the world every year and ensuring that potential readers can obtain those items they specifically require is an indispensable task, which will have to continue to be fulfilled by the booksellers.

If anything, the booksellers are taking advantage of the new possibilities opened up by computers and telecommunications networks, the Danish project described by Bjarne Hjort being a good example of this, which is being followed in other European countries, notably the Federal Republic of Germany and the United Kingdom. What is important today, therefore, is that booksellers should be made fully aware of these developments so as to use them to the full, for their own benefit and for that of their customers. It was recommended that steps should be taken to do so, for instance by organizing appropriate training courses, especially for young people entering the bookselling profession.

The librarian

The chairman quoted a statement he had heard recently at a conference held in the United States "The library is dead. Long live the librarian'." This is presumably a way of saying that, in the future, if most information is published electronically, the librarian will have no bookshelves to keep in order. He will however be expert at browsing through electronic files and will, with the help of his terminal, direct people to the information they require.

This scenario was not rejected out of hand. Perhaps it will in due course apply in certain special contexts, say in research laboratories. The majority of local libraries will clearly not be blotted out of the landscape any more than local booksellers.

The publisher

The advent of new technologies is generally regarded today by publishers more as a threat to their livelihood than as an opportunity for a better and more exciting future. This attitude is not based on a sheer reluctance to accept change but on the genuinely unfair impact certain technologies are having on the profitability of their operations.

If it is true that publishers fulfil a socially useful role, as well as the authors whose works they publish (no one seems to be prepared to dispute this), then they are entitled to a fair reward. The normal way in which they reap this reward is through sales of copies of their publications. However, copying equipment has become so efficient, from the point of view of quality as well as cost, today, that it is only too easy to produce further copies from a single "master" which has been duly purchased.

There admittedly exists legislation, in most countries, to protect authors and publishers against the effects of copying. In some countries it is totally prohibited, in others it is allowed under certain conditions only, e.g. as long as royalties are paid.

Unfortunately, it is extremely difficult to catch red-handed those who infringe copyright law and thus take action against them, unless they engage in the production of large numbers of copies for resale. In any case, in several countries, copying for private research purposes has been made legal, it being assumed that this contributes to the general spread of knowledge.

The publishers' view is that this assumption is misguided. The general spread of information is certainly socially desirable, but allowing copyright to be infringed is a self-defeating way of promoting it because, in the long run, publishers will tend to go out of business, with the result that the flow of information will be seriously reduced at source and there will be fewer people engaged on checking and improving its quality and making sure it is properly marketed.

At the moment, it is the world of scholarly journal publishing that is most concerned about this copyright issue. Not only has it become, technically, so easy to make copies, but the growing efficiency of secondary services, due to the use of computers and telecommunications networks, is tending to make it easier for users to identify individual articles of potential interest to them and to secure photocopies through the library system.

It could however be said that the publishers of the written word have worse things coming to them. There is no evidence, for instance, that sales of general interest magazines or books are being affected by the availability of copying technology. Will they eventually be affected as much as the publishers of music on records and cassettes today?

The average user of information is of course quite unmoved by, or rather totally unaware of the problem faced by publishers. He just wants information. The librarian, whose *raison d'être* is to provide access to literature and who lives in a traditionally non-profit-making environment, is principally interested in satisfying the user's needs as quickly and cheaply as possible.

Thus librarians are instinctively not well disposed towards procedures which in any way make copying more difficult or expensive.

A representative of the profession, responsible for a large university library which runs a regional inter-library loan scheme, quoted the following figures: in a year, about 6 000 transactions (in the form of loans or photocopies) take place; this is to be set against a total of 8 000 subscriptions to journals taken out in the area. Can publishers really claim that their interests are being seriously damaged?

A representative of the publishing industry retorted that such statistics are basically irrelevant: it is still wrong, as a matter of principle, to copy. In any case, it must not be assumed that an inter-library loan involves no costs; in one country of the EC, at least, they are estimated at 12-15 e.u.a.* (The rapporteur would like to comment on this: the cost of ordering, with or without using the services of a bookseller, is also substantial, not to mention the excessively long delays that may be involved.)

The librarians present did not fundamentally question the right of publishers to charge royalties. Whatever system is adopted for this purpose, they expressed their preference for one that would be simple, and therefore cheap, to administer. Moreover, they pleaded in favour of a moderate level of royalties, in the interests of everyone concerned. Indeed, ironically enough, if royalty payments absorbed an excessively high proportion of their budgets, they would have to cancel some subscriptions and rely even more on genuine inter-library loans.

In the view of the rapporteur, this battle between publishers and librarians (or skirmish, rather; as was said above, the publishers have not really been hit hard yet) has up to now led to a complete stalemate - and fortunately so, because it is difficult to perceive serious grounds for a battle in the first place. Both parties, after all, share a common basic ideal, that of communicating information efficiently (information being here used in its widest sense, to cover not only data and knowledge but also culture, entertainment, etc.). They do this in slightly different ways. Surely in the society of the future, which some nickname the " information society", because of the expanding role which the generation, storage, transfer and processing (mechanical and especially intellectual) of information will play within it, it should be possible for both of them to find comfortable niches.

* *1 e.u.a. is roughly equivalent to £0.65*

What are the publishers really saying? They are accusing the librarians of indulging in a form of "on-demand publishing". This is not a mere demarcation dispute because, as we have seen, the activities of the librarians smack of piracy.

By the way, it is worth stressing that the dispute is opposing only a minority of publishers against a minority of librarians. The average local, institutional or company librarian is a very good customer indeed of the publishing industry. The real problem arises when certain public libraries are set up on a sufficiently large scale to be able, thanks to their comprehensive holdings, notably of journals, to provide a nation-wide (at least) loan and copying service. As long as the efficiency of these large libraries remains moderate, the publishers do not really mind, but when they start being able to satisfy requests for photocopies of articles within a few days, they object.

Surely there must be a way to resolve this paradox without damaging the interests of the user, who is inevitably in favour of the fast service that he is receiving from certain large public libraries.

The key to the success of these libraries is the comprehensive nature of their holdings. This makes life extremely simple for the user, who just has to remember one address, apart from the fact that the delivery mechanism involves no intermediaries and is thus potentially extremely rapid.

There is yet another favourable point, of course: the service is cheap, if only because libraries of this sort receive a state subsidy and need not recover the total cost of their operations. In the view of the rapporteur, this is however not the main point. Experience has shown that raising charges does not affect demand substantially. On the other hand, a factor such as reputation for speed of delivery stimulates it.

The rapporteur would therefore suggest to the publishers that their best method of defence is perhaps attack. If the kind of service provided by the large public libraries meets a real and expanding need and is in reality a form of publishing, then it could be argued that it is up to the publishers themselves to apply their expertise, their spirit of enterprise and their resources to this task. Implementing such an idea would involve setting up document delivery centres, the critical size of which, if they are to become economically viable, is likely to be very large indeed, probably too large to make it possible for any single publisher to envisage taking upon himself the whole risk of such an enterprise. Perhaps the answer lies in the creation of joint document delivery subsidiaries.

These delivery centres would have to remain adaptable to new technologies, as and when they become available and useful. They could, for instance, become clearing-houses for electronic journals.

It is to be hoped that they would, in due course, arrive at a high degree of efficiency. In the meantime, the large public libraries will have to continue, to the best of their ability, to satisfy users' needs. (As they are dependent on public funding, they are not always actually in a position to provide truly satisfactory services; in some countries, they are non-existent.) Some kind of royalty system will have to be devised in order to cover publishers' interests during this interim period.

The user

End-users of information tend to be the last people to be consulted when information policy matters are discussed. Several participants pleaded in their favour and regretted that they were not represented.

Others pointed out that, in their experience, users are very elusive people: there are so many of them, all with different needs and tastes. Isn't it precisely the role of publishers, booksellers and librarians to be their brokers?

A librarian made the following point. In his opinion, the present electronic information retrieval systems are rather costly and are relatively little used because they are not really user-friendly. The briefcase computer, will in his opinion, bring about a much-needed democratization, because each user will be able to "tank up" with a fairly large amount of information tailored to his individual needs.

A new approach to dissemination of information towards the developing countries

Several participants pointed out that the developing countries are even more in need of information than the developed countries and suggested that every effort should be made to use the new technologies in such a way as to improve the transfer of knowledge towards them.

At first sight, the new technologies take the form of expensive gadgetry and the outlook is therefore not promising.

It was argued that the problem had to be tackled with more imagination. The communication of knowledge is traditionally associated with the written word, which means that the potential recipient must be literate. Now it is a fact that about 80% of the world's population today is illiterate. Would it not be useful to probe deeply into the various possibilities opened up by the substitution of electronic for "alphabetic" communication?

This is a huge subject, which it was generally agreed deserved detailed study.

Conclusions

Chairman and rapporteur: J. Michel GIBB

The advent of new technologies, in any branch of human activity, can lead broadly speaking to developments of two kinds. It can make it possible to satisfy conventional needs more economically or it can generate totally new products or services.

The publishing industry being no exception, it is being affected in both these respects. The changes which the production of daily newspapers is undergoing at the moment are typical of the first kind of development. Although an important and sometimes painful revolution is in progress, dictated by the urge to bring down costs, the average reader is hardly conscious of it, because a daily newspaper looks very much today as it did half a century ago, and there is no reason to suppose that its physical appearance will be substantially altered in the next fifty years.

Videotex is an example of the second kind of phenomenon. It will soon be putting at the tip of millions of people's fingers, literally, access to information of all sorts, from the latest football results to an inventory of human knowledge, in a totally new form. To some extent videotex services are just competing with traditional information carriers. You can find football results in your newspaper, hear them on the radio or even obtain them on the telephone. However, the sheer convenience of being able to summon them onto the screen of your home TV set by depressing a few keys on a simple keypad, at any time of the day or night, may prove appealing. In other instances, videotex will be opening up entirely new possibilities, like consulting a wide range of constantly updated encyclopaedias and directories. It is in fact too early today to guess at all of these possibilities. It is precisely up to the publishers to imagine them and offer them to the public.

Very little was said during the symposium about the first kind of development. There was a general impression that many of the products of the publishing industry would continue to be offered in their present form, not only newspapers and a large number of magazines and journals, but books, manuals and the like. This would be particularly true of the wide range of publications which have to do with general culture and entertainment. Better and more economic printing and distribution methods would of course constantly be sought.

As the majority of participants, whether authors, editors, booksellers, publishers, librarians or documentalists, came from the world of specialized information transfer, particularly in the scientific and technical fields, the discussion concentrated on the effects the new technologies would have, for good or for bad, on their own working environment.

The major technologies involved constitute a set of tools, mainly electronic, which can be used for a variety of purposes under the following three main headings:

- generation and processing of information;
- storage and retrieval of information;
- copying and transmission.

The traditional medium for the transfer of specialized information is the journal. It was generally felt that the average author was quite happy with this system, although he might well criticize its slowness.

On the other hand, the cost of editing, typesetting, printing and distributing journals is steadily increasing in real terms. A large percentagg of this cost does not vary with the number of copies distributed. Thus the very specialized journals, with small circulations, are being seriously affected. If they put up their subscription price to compensate for this, they run the well-known risk of making their situation even worse, as librarians, whose budgets are not elastic, will be faced with the need to cancel subscriptions and will tend to bring the hatchet down on the more specialized journals.

For some time already, publishers have been trying to alleviate this difficulty. The "camera-ready journal", produced from the authors' original manuscripts, is one of the obvious products of their efforts. It does away completely with typesetting or, more accurately, it shifts the cost of text-composition to the author.

One of the drawbacks of this system is that the whole process of bringing an article up to standard, through the work of the editor and his referees, is made more cumbersome, unless access to word processing equipment can become commonplace.

In a situation where authors, referees, editors, publishers, etc. are usually in different locations, the spread of such equipment opens up exciting possibilities, especially in the era of "telematics", which is just beginning. The availability of telecommunications networks is full of promise because it brings the various partners virtually together, thus not only saving time but also improving scope for interactive communications, which should lead to better quality literature. It is clear, in this context, that communicating word processors must be cheap enough to make it possible for small institutions to acquire them and be intercompatible, whoever the manufacturer of the equipment is.

Standardization, at a European level at least, is necessary if this objective is to be achieved. This was stressed several times during the symposium. Those who have had anything to do with devising standards, and especially enforcing their implementation, know that the task is often frustrating and always exhausting. Yet it must be carried out or else the kind of future which has just been outlined will remain a dream.

The fear was expressed that the introduction of the new electronic devices could bring in its wake a general decrease in the quality of primary information. This fear was evidently not shared by all participants some of whom claimed on the contrary, as suggested above, that quality would improve if their full potential were used. This is perhaps not a true contradiction. Admittedly there may come a day when it will be technically very easy indeed for individuals to disseminate "electronic documents" at will. Hence it is probably true that a good deal of unchecked material, of which a high proportion may be of dubious quality, will be available through electronic networks, but surely it will be possible for the user to pick out without trouble, in this mass of material, the articles published by reputable electronic journals. In any case, if authors, editors and referees find that their activities can be streamlined, thanks notably to communicating word processors, this does not mean, as in the case of newspapers, that the final product need be issued in an unconventional form. The final product can be a printed journal, produced from camera-ready typescripts or, better still, at little additional expense, from automatically photocomposed texts.

More fundamentally, are electronic journals required in the first place? Although there is an observable trend towards the transfer of the task and expense of composing texts towards the authors, the costs of printing on paper and of distribution will continue to weigh more and more heavily on the small-circulation journal. No participant disputed this. But if an alternative to the conventional journal must be sought, why must it be an electronic one?

The question was not really put point-blank in this way during the meeting. As a result, it seemed to be assumed that there was in fact no other option. And yet other better-established cost-saving techniques are available, such as microreproduction. However, microfilm and microfiche appear to be firmly associated with documentation in its archival rather than informational function. Users may be prepared to consult microform publications occasionally, but they do not seem to want actually to read them.

Will the future of electronic publications be brighter than that of micropublications? Again, it was just assumed that this would be the case. One can, of course, think of reasons why it should be. One of them is the far greater flexibility and speed of access offered by the electronic environment. It is here that another consideration was brought forward, with all the trappings, initially, of a red herring. I mean, here, the question of synopsis publishing.

The synopsis journal was proposed several years ago as a possible solution to the difficulties specialized periodicals are encountering in making ends meet. The full-length archival version of an article is made available via a cheaper medium, e.g. in microform, whereas only the short version, which contains only essential information, is printed and disseminated in the conventional way on paper.

This concept, which makes such economic good sense in theory, has not met with much success, perhaps mainly on account of author-resistance. A marriage was therefore proposed between synopsis publishing and the electronic medium to give an electronic synopsis journal. As someone pointed out, it might be refreshing to find real information being published over networks like Euronet rather than just bibliographic data and mere abstracts.

Several participants expressed their hankering for a better understanding of the whole primary information transfer process. They did not wish the subject to be analyzed from one point of view only, but from that of all the partners involved, which was completely in line with the spirit of the symposium. Publishers may find it financially advantageous, for instance, to encourage authors to take over responsibility for text composition, but this does not eliminate costs; it just shifts them from one group of people to another. What is wanted is a study showing clearly what different tasks are involved and analyzing the options for each one of them, as to who should carry them out and what methods should be used, from a cost/benefit point of view.

It became clear in the course of discussion that most of the partners in the information transfer chain would tend to benefit rather than suffer from the introduction of new technologies. This is true of authors, editors and referees, who should be able to work faster and more effectively. It is true of booksellers, who should be able to provide a better and quicker service. It is true of documentalists, information officers and information brokers, whose role will become more important because of the increasing complexity of the information scene.

The publishers, too, stand to gain insofar as some of the new technologies can enable them to issue their conventional products more economically. There is statistical evidence to the effect that in the countries of the European Community publishing houses are today on the whole increasing rather than reducing their establishments. However, the widespread availability of efficient photocopying equipment (to be followed up soon by yet other copying technologies) is placing them before a serious problem.

As was seen during Session V, the enforcement of copyright law is necessary to protect their interests, but is not a sufficient measure in itself. It has been suggested that, inasmuch as the danger comes mainly from a few libraries with comprehensive holdings providing a fairly fast document delivery service, it is up to the publishers themselves to create document delivery establishments of their own, which could apply the most sophisticated technologies when necessary. Such establishments need to be organized on a large scale if they are to have a chance of success. As the publishing industry is made of medium to small enterprices, action may have to be taken collectively, in which case the European Communities could provide help, say in the shape of loans and even outright grants.

It is fortunate, in a very fundamental sense, that the publishing industry should be made up of so many independent units. Diversity of sound information sources is a key to the health of a democratic society. Thus, if it is true that one of the effects of a realistic approach to the application of new technologies in publishing is the need for large-scale investment, it is to be hoped that it will not be left to other branches of industry, the computer industry, for instance, to fulfil it, but that the publishers will rise to the challenge themselves.

Issue file

compiled by
Lydia GERLACH

Presented as a basis for the PANEL DISCUSSIONS - SESSIONS II to V

The *issues* and the different *positions* related to each issue which are contained in this *issue file* have been compiled out of hundreds of issues and positions suggested by the panel members prior to the symposium. The file was constituted on the basis of their suggestions and presented to all participants of the symposium, so that they could write, during the symposium, their *arguments* in relation to the stated *positions*.

After the symposium, these *arguments* were added to the relevant *positions*, and they are now included in this revised version of the *issue file*.

(As the original issue file submitted to each participant was written in English, the collected *arguments* have also been translated into English; we hope that in this case, the use of a single language will simplify comprehension, because the subjects treated are somewhat controversial, and translation into the different languages, not to mention the very personal jargons that have been used, could have favoured misinterpretations).

This *issue file* served as a test as to whether it was possible to have a ''*paper-discussion*'' going on in parallel with the spoken conference, so that in spite of the limited time and a great number of participants everybody would have a chance to have his say.

We hope that this first try will be well accepted by all participants, whom we thank for their cooperation.

List of issues

No. 1: Will the new technologies have an impact on the health of the personnel involved: if so, what solutions are suggested.

No. 2: What impact will the new technologies have on the general job structure in the publishing sector, and what attitude will allow us to cope best with the situation and/or permit the most appropriate restructuration?

No. 3: What are the reasons for users' reluctance?

No. 4: How can the users' reluctance to accept new publishing technologies be overcome?
- Education
- Improvement of existing methods.

No. 5: The individual publisher's first approach to new publishing technologies:
- Whether or not to invest?
- New business ventures
- How could government institutions help?

No. 6: First organizational impact of adopted new publishing technologies on the publisher, the printer, the compositor, the editor and the author, and the modification of their relative roles.

No. 7: Review of new publishing technologies which are already used for processing, production and data storage, their advantages and disadvantges, and suggestions for improvement.

No. 8: Review of new forms of output, communication and dissemination media, their advantages, disadvantages, and suggestions for improvement of technologies, methods and policies.

No. 9: Journals and newspapers confronted with electronic media.
How can government institutions ensure a good overall access to electronic publishing?

No. 10: How can libraries, bookshops and secondary services use the new technologies and how can they cooperate with publishers?

No. 11: Legal implications of new publishing technologies.

ISSUE No. I:

WILL THE NEW TECHNOLOGIES HAVE AN IMPACT ON THE HEALTH OF THE PERSONNEL INVOLVED: IF SO, WHAT SOLUTIONS ARE SUGGESTED

POSITIONS

No. 1 - *Perpetual reading and manipulation via television screens, for example, appears tiresome for the eyes. This is why personnel involved in such activities should work only a limited number of hours.*

ARGUMENTS:

1.1 - Most VDU operations in *word processing*, and other forms of text and data processing, require far more time for looking at source documents than at screens. Better task lighting is necessary. Many people using screens are women, who have, before using VDU's, refused to admit to themselves that their eyesight needed correction.

1.2 - These activities do not only appear but *are* tiresome. Improvements could be obtained by researching for the least tiresome forms of display. Consider as an example for the suggested limitation of working time that simultaneous interpreters are also granted working time arrangements corresponding to their (exerting) activities.

1.3 - I realise the temptations to use health questions as a "defence" against the new technologies. However, in my opinion, the health hazards must be seriously investigated. As a scientist I have been asked to look into the effect of long-wave electro-magnetic radiation on membrane functions. Even though I did not have the opportunity to take up the line - if connections of this type occur, they may lead to biological disturbances of medical consequence. Furthermore, my son, who is doing military service watching a screen, is allowed to do so only on a halftime basis.

No. 2 - *It is necessary to provide good medical investigations on a large scale in order to have realistic data available on how far health could be affected through the new technologies. It is however necessary that these investigations be executed by a completely neutral body (neither industry, nor union-orientated medical doctors)*

ARGUMENTS:

2.1 - Impartial *medical* investigations are the only way to detect possible detrimental effects. If necessary, measures described in *position* No. 3 will have to be taken. *Psychological* investigations into the effects of these technologies should be carried out as well (rupture with the familiar way of working).

2.2 - The detriment to the employee's health through technological evolution may result from two factors:

1) OBJECTIVE: The appeal of an economically attractive market requires considerable technological efforts from engineers in order to beat competition. Therefore, problems related to the *utilization* of the equipment concerned are neglected (e.g. tiredness, monotony, etc.).

2) SUBJECTIVE: Technical progress faces a latent refusal in the western countries (unfavorable environment, ecology, unemployment). This involves psychological reticences which will lead to physiological outlets.

SOLUTION: These problems should *not* be neglected. At the moment when a technological innovation is realised, its utilisation ought to be investigated by pluridisciplinary teams (engineers, designers, psychologists, ergonomists, physicians, trade unionists) in order to achieve the adaptation of equipment to man.

No. 3 - *It is necessary to investigate the technical reasons how the health of the user can be affected through new technologies, in order to improve the design and elaborate technical rules corresponding to health standdards.*

ARGUMENTS:

3.1 - Adapting machines to man and not vice versa: Datasaab, for example, has put theory into practice with designs tailored to human needs. Their brochure 'Ergonomics - The Third Factor' illustrates potential problem areas in a screen-based installation and offers guidance on how to avoid them.

3.2 - With the aim of designing a sound working environment Skandia, the large Swedish insurance company, carried out a survey among computer operators in 1976, in order to determine the discomfort they experienced. The percentage distribution of discomfort revealed by the survey was as followed: Eyes - 54.8 %; Head - 30.3 %; Neck - 15.2 %; Back - 43.7 %; Shoulder - 25.1 %; Wrist - 18.8 %.

3.3 - At Nottingham University, a computerised man is providing the vital human element in computer aided design. "Sammie" (System Advancing Man/Machine Inter-active Evaluation) can, on a giant display screen, be made to simulate any posture or movement that a human being might adopt in a given working situation. Thus, equipment is designed interactively with the 'body' in the fullest ergonomic sense.

3.4 - The following *'WORK STATION CHECKLIST'* has been produced in *INDEX* November 1979' (p. 23) with the permission of Datasaab Ltd.:

Work station checklist

External environment
- Screened windows
- Non-glare overhead lighting
- Walls not too light
- Balanced contrasts
- Subdued overhead lighting
- Comfortable temperature
- Low noise level

Work station
- Placed to meet individual needs
- Varying work
- Stimulating tasks
- Breaks

Console
- Movable
- Inclinable
- Low height
- Clear symbols
- Not too light in colour

Screen
- Reflex-free
- Restful picture for eyes
- Well focused symbols
- Variable light intensity
- Separately adjustable caption brightness
- Hood

Chair
- Adjustable height
- Adjustable backrest
- Restful seat
- Armrests where necessary

Table
- Adjustable height
- Large enough working surface
- Individual placing possible
- Space for sunken console
- Document stand

ISSUE No. 2:

WHAT IMPACT WILL THE NEW TECHNOLOGIES HAVE ON THE GENERAL JOB STRUCTURE IN THE PUBLISHING SECTOR, AND WHAT ATTITUDE WILL ALLOW US TO COPE BEST WITH THE SITUATION AND/OR PERMIT THE MOST APPROPRIATE RESTRUCTURATION?

POSITIONS

No. 1 - *Large areas (of the printing industry for example) will disappear in terms of employment, skill, turnover and capital deployment and will cause important social and commercial restructuring.*

ARGUMENTS:

1.1 - It is of national interest to create the psychological premises for these inevitable restructurations. "Learning womb-to-tomb" is the answer. The social partners ought to arrange for adequate structures allowing continuous education and, where appropriate, educational leave.

1.2 - A study of past "upheavals" in employment fluctuations, change, decline, consequences etc. may prove a convincing test of the various and opposed positions of this Issue (applies to all positions of this issue).

No. 2 - *Only highly qualified experts will occupy interesting jobs. All other jobs will become more monotonous and less satisfactory to the individual.*

ARGUMENT:

2.1 - He who is more efficient will and should have certain advantages. This prevails at present as well as in the future.

No. 3 - *Though certain jobs will disappear through new technologies in the immediate future, the same technologies will create at least an equal number of new and different jobs in the medium and long term.*

ARGUMENTS:

3.1 - The Commission should give *urgent* attention to:

a) Investigating new jobs and skills likely to be created/needed by the office and publishing automation revolution.

b) Studying new possibilities of application of the new technologies which could create profit-earning jobs.

c) Gearing/assisting EEC suppliers to meet the demand for new technology, rather than relying on US products as a substitute.

No. 4 - *A smooth passage to new technologies within the individual enterprise will give to the temporarily unoccupied personnel sufficient time for re-education and allow an appropriate restructuration.*

ARGUMENT:

4.1 - The enterprise investing in new technologies can hardly afford to pay at the same time for temporarily unoccupied personnel. Financial assistance should be provided in these cases by means of arrangements between the social partners and the government

No. 5 - *The individual enterprise could best cope with the necessary job restructurations by minimising recruitment and increasing wages and salaries with the addition of bonus schemes and incentive payments.*

ARGUMENT:

5.1 - This hypothesis should end after the word "restucturation". Which measures would be the most appropriate should be left open. It will depend on the economical efficiency of the overall operation after implementation of the new technologies. This economic efficiency is mostly described too optimistically by the producers of the equipment concerned. In certain cases, there will be no savings of personnel at all.

No. 6 - *There is no problem at all - the natural interaction of free economy will re-establish the equilibrium of enterprises and jobs.*

ARGUMENT:

6.1 - The mechanisms of free economy are not that perfect. There are viscosities and delays which cause a good deal of technological unemployment and may lead after a certain time to important human problems.

No. 7 - *Constitutions should include the "right to work"*

ARGUMENTS:

7.1 - Yes, *but*

a) not the right to do useless work;

b) also the protection against being excluded from work by union activity (i.e. "right to work" in the full *American* sense).

7.2 - Constitutions with the "right to work" seem to degenerate to a "right to be present but not to work" or for "three people to do the work of one". What is monotonous to one person is not necessarily monotonous to another. To rely on natural wastage or rather artificial means may not provide openings for people who try to find jobs and are able and willing to do them properly.

No. 8 - *A kind of "tripartie" (standing consultation between employers, employees and governments) could continuously review the economic situation in the publishing sector and seize all possibilities of job creation through the new technologies, eventual conversion of enterprises and jobs, means for re-education and possibly financial support to those who are temporarily weakened by the advent of new technologies.*

ARGUMENT:

8.1 - The insertion of new technologies must be prepared beforehand on the following basis:

- statistics of employment;
- education to cope with necessary evolutions;
- education for other jobs which are open in the same sector;
- education for other jobs in sectors with related activities.

ISSUE No. 3:

WHAT ARE THE REASONS FOR USERS' RELUCTANCE?

POSITIONS

No. 1 - *Some users are basically reluctant because they believe that the increase of productivity and wealth obtained through new technologies has never contributed to the happiness, satisfaction or freedom of the individual.*

ARGUMENTS

1.1 - This is the classical position of the French professionals involving a complete cut between the publisher and the printer and, consequently, leading to the preservation of conventional methods.

1.2 - As a user, I do not feel reluctant to accept new technologies if they give rise to more comprehensive, quicker and cheaper information.

No. 2 - *Some users are reluctant because they fear to lose their freedom in the choice of knowledge and information, as new technologies may lead to economic concentrations and monopoly situations.*

ARGUMENT:

2.1 - There are many users in the world whose choice of information possibilities is extremely restricted. If the new technologies are going to expand their mostly politically determined circle, these users will not have a chance to improve their situation.

No. 3 - *Some users are reluctant to accept non-book publications because they cannot have access to the desired knowledge at any moment or any place they want to, due to the need for interface equipment, sometimes occupied telecommunication lines, the dependence on electricity, and because the information stored through new technologies may be subject to technical defects.*

ARGUMENTS:

3.1 - The user does not seem to want the present technological evolution. In this context I confirm observations made at the Jülich Research Centre. The user's fear regarding a technology which he cannot master and his mistrust towards intermediaries (documentalists) seem to be the causes for the present disaffection. The producers of the present technologies seem to ignore the education and wishes of the user. They create more and more sophisticated systems, the keys of access to which become increasingly difficult to discover.

3.2 - The reluctance on the part of scholars to use the new technology (terminals, microform) has been brought about, to a large extent, by the poor quality of the existing systems and equipment. A data base search today can throw up a mass of ill-digested material which distracts the reader. Equally, badly designed equipment (e.g. micro viewers, terminals) actively discourage use.

No. 4 - *Some users are reluctant because when searching through on-line systems they waste time browsing through information that is already obsolete but still stored.*

ARGUMENT:

4.1 - The question of deleting obsolete and superfluous data from the storage unit becomes more and more important and corresponds in some way to the problem of decommissioning nuclear power plants. However, this involves the dangerous possibility of introducing censorship, at least of giving priority to literature for other reasons than those concerning their real informative value.

No. 5 - *Some scholars are reluctant to accept non-book publications because they cannot browse through several scientific works at the same time and therefore lose their capacity for simultaneous comparison.*

ARGUMENTS:

5.1 - This is why the new technologies should be complementary and not exclusive.

5.2 - Copying methods could give a solution to this problem; this is why practical legal solutions concerning copyright become increasingly urgent at the same time.

No. 6 - *Some users, particularly families that are passionate readers used to sitting together in a group, each one reading a different book and from time to time reading aloud to each other, fear that through the use of non-book publications they might have to give up this habit or agree that they should all read the same book at the same time.*

ARGUMENTS:

6.1 - This idyllic situation does not fit into the kind of literature to which the new technologies are applicable.

6.2 - This applies to reading for relaxation, not to assimilating information.

No. 7 - *Some users are reluctant because they consider knowledge transmitted through a non-book medium too persuasive and demanding too rapid assimilation, and that therefore their thinking will be lagging behind.*

No. 8 - *Automated systems themselves frequently make mistakes but they place a demand for total accuracy on the user in carrying out procedures and pressing the right button.*

ARGUMENTS:

8.1 - If the correct button is pushed, the right answer will be issued. However, the user cannot check whether the machine itself works correctly. Users may also waste a lot of time in front of screens; therefore, they ought to be helped for a long period of time by intermediaries who sufficiently understand the subject matter *and* the technological means used.

8.2 - Users are reluctant because the new technologies are not well developed/adapted to *serve the users.* On the contrary, it often seems that the user has to serve the new technology, and this is a perverse situation.

No. 9 - *Some authors are reluctant because, when using, for example, word processors directly during the creation of their work, they lose their spontaneity and intimacy of expression, in having to concentrate simultaneously on a format, a keyboard and looking at a screen.*

ARGUMENTS:

9.1 - Maybe, but most authors have no access to word processors, particularly not during out-of-office time when most work is done that really requires intense concentration and absence of disturbance.

9.2 - Anyone who can look at work, screen and keyboard at the same time is a physiological freak; neither can he *concentrate* mentally either. Why is this seen as so different from using a typewriter and a source document? Anyone who has seriously used a VDU *knows* it is instinctive after a short while and the freedom to correct enhances the ability to create.

9.3 - For authors who communicate research results, *word processing* is a very important instrument, because it allows the addition of new experiments, text modifications and preparation of the final text version without completely rewriting each time.

9.4 - It has been observed in at least one German daily newspaper office that the vocabulary of editors/journalists who have to enter their contributions through new input media (terminals) is diminishing.

9.5 - Six months experience of a word processing installation at NRCd reveals that far from losing spontaneity and intimacy, authors appreciate word processors for possessing exactly these virtues. The ease of correction, formatting and the action of the keyboard (much improved over typical typewriters) greatly assist authors in their work. Most are extremely reluctant to revert to more traditional methods of input.

No. 10 - *Some authors are reluctant to use new technologies because standardization is lacking and they have to adapt themselves constantly to new formats and systems.*

ARGUMENT:

10.1 - Because of the distribution of information over several data bases where each one requires a specific method of interrogation and different keys, the user is obliged to go through different apprenticeships, which he refuses: "no time!". Thus he has to be helped by a professional searcher.

No. 11 - *Many users are simply reluctant to accept new technologies because they are not sufficiently aware of the usage they can make of them and the advantages and possibilities they can bring.*

ARGUMENT:

11.1 - I consider this the basic reason for user's reluctance. Most other positions can simply be considered as "opinions" because of user's lack of understanding and the present premature use of technology.

No. 12 - *The user group, consisting of potential investors in new technologies and the staff directly concerned with editing or publishing through the new media, is partly reluctant to accept the new technologies because current information on them, coherent rules and legal adaptations are still missing.*

ARGUMENT:

12.1 - Yes, and the information that is available is not sufficiently practical and is too confusing. Too much sales publicity, too little information.

ISSUE No. 4:

HOW CAN THE USERS' RELUCTANCE TO ACCEPT NEW PUBLISHING TECHNOLOGIES BE OVERCOME?
- EDUCATION
- IMPROVEMENT OF EXISTING METHODS

POSITIONS

No. 1 - *Regular consultation with associations in the field of information and publishing should be arranged in order to establish educational programmes for different categories of users.*

ARGUMENTS:

1.1 - If the new technologies led to a situation that renders getting informed more difficult for the users and where they had to spend excessive amounts of time for learning, it would not be worthwhile to introduce these technologies.

1.2 - In the first instance very elementary courses are required for the "over 35" generation who are working *now* and are responsible for policy decisions.

No. 2 - *Permanent education of children and adults could include practical courses on how they could instruct and inform themselves better and more quickly through the use of the new information media.*

ARGUMENTS:

2.1 - Experience of a publisher specializing in educational material: Our final consumer is the teacher (and his students) to whom we furnish pedagogical tools. We have noticed that our user (the pedagogue) is not interested in adopting educational tools realized by new technologies; paper remains their medium of choice. This does not mean that the world of education is closed towards novelty. There are experiments on teaching by means of computers (e.g. mathematics) and integrated television circuits (e.g. at the pilot Lycée of Sèvres or in Vincennes).

The French Minister of Education has expressed during a congress on "methods of learning how to read", that 6-year-old children should learn how to read on a terminal, thus opening schools towards the modern environment. But the scholastic publisher's place is still in the production of "traditional" working tools. We are, however, open to evolution; there is no question of blocking it but of keeping one's ears open: hence my presence at this seminar.

No. 3 - *To educate future authors, university education in all disciplines should include an introduction to new publishing technologies.*

ARGUMENTS:

3.1 - Yes, but do not limit education to new publishing techniques. The entire subject of "information use and transfer" should be taught as a basic discipline comparable to "energy and its generation, use and conservation".

3.2 - Most scientists produce one or two papers in their lifetime, some can produce as many as 200 or more. In his lifetime such a scientist will read, however, many many more. It is essential to have immediate access to journals which are publishing papers in one's own field. As a far greater amount of literature can be searched through modern systems each scientist should learn to take advantage of them.

3.3 - (concerns also position 4)

One cannot request the author to prepare his texts input-ready for modern typesetting machines. The publisher should offer the author different possibilities of text transmission (e.g. paper, magnetic tape, floppy disc, etc.). Therefore, future authors must *not* be familiarized already in school with modern publishing media, the more so as these methods are continually being improved and modified.

3.4 - Future authors must be informed about the new publication technologies already during their education. But the main task of these authors in general is performing technical and scientific research; authors are not there to adapt themselves to the decisions of the information industry. When this industry decides to introduce new technologies these technologies have to be "author-friendly". Without authors, the information-industry ceases to exist.

3.5 - Authors, editors and referees should not only be informed about the possibilities of new publication technologies but also learn how to communicate efficiently via present paper-based technologies which serve as a communication basis today. This education will be applicable also to new technologies; just some supplementary instruction will have to be added according to the medium used.

It is also necessary to explore the possibilities of rapid interaction between authors, editors and referees (often slow and inefficient at present).

No. 4 - *Management education in universities should include subjects like: word processing, data processing, image processing, information storage and retrieval, as well as behavioural and organizational aspects related to the application of these techniques.*

ARGUMENT:

4.1 - The average student ought to be spared from such ballast. Competent information services should be organized in such a way as to communicate to *everyone* the required information as quickly, correctly, simply and cheaply as possible.

No. 5 - *More jobs for advisor analysts could be programmed so that they could contact a larger public in order to facilitate the communication interface and the coordination of functioning systems.*

ARGUMENT:

5.1 - The word "advisor analyst" sounds complicated. I suppose that "information broker" is meant. This profession will play an important part for a long time to come. It should be possible to refer to these intermediaries in libraries and at other appropriate places.

No. 6 - *Libraries could give their clients training and refresher courses on ST information transfer with particular attention to the newest technologies.*

ARGUMENTS:

6.1 - Such courses will not be required if libraries would employ a sufficient number of information intermediaries (information brokers). Students and scientists should not spend their time learning complex technologies and their tricks, but know how to assimilate critically the information received.

6.2 - Libraries could do more - they could make their equipment available to publishers and thus accelerate the information flow.

6.3 - The multiplicity of sources with different kinds of access will exclude a real efficient training. This is why a common methodology and programming of access channels and command codes is required first.

No. 7 - *Courses on new publishing technologies and their use could be stored on video cassettes and made available to the public through schools, libraries and the book trade.*

No. 8 - *Even courses on traditional ways of publishing (history of printing, etching, etc.) containing good picture material could be stored and distributed with the help of new media (microfilm, video, etc.) so that the user receives at the same time an all-round image of the appropriateness of the different publication methods to the kind of information communicated.*

ARGUMENT:

8.1 - Who would pay for this and why? Manuals for these technologies do already exist for those who are interested. Supplementary manuals will be added for new technologies to come.

No. 9 - *More easily understandable information about new technologies could be directed to specific target audiences of authors and information users (e.g. editorials in learned society journals, etc.).*

ARGUMENTS:

9.1 - This can be done for mere information purposes. New media and technologies are, however, meaningful only if they give easier and sure access to information - not if they complicate this access and presume new and remote areas of knowledge.

9.2 - Such information must be practical and include some indication of the level of costs (does it cost £1 000/£10 000 or £1 000 000?).

No. 10 - *Policies in the employment of research personnel should be such as to relieve authors from the "publish or perish" syndrome.*

ARGUMENTS:

10.1 - If only it were possible! Observing the rubbish produced in universities in order to re-invent again and again new titles for doctoral theses, one realizes that an old tradition is perpetuated which represents pure nonsense, to the students and to the scientists involved. But who will ring the bell?

10.2 - Any attempt to destroy the existing system whereby research scientists are assessed by their publications output would be fatal to the publishing industry. It is probably the only system that works provided that the peer review system is maintained. Thus we should strive to ensure that quality is maintained so that the present system preserves its credibility.

No. 11 - *Leading scientists should be persuaded to be open to the use of new publishing technologies, so that new media stop being considered as "second class" because only relatively young authors use them.*

ARGUMENT:

11.1 - What do leading scientists have to do with this question? The new technologies must prevail because of their economic and practical features and not because of a publicity action by scientists who have been misused for this purpose.

No. 12 - *In cases where governments have a direct influence on the publication of research information, they should encourage the use of new media.*

ARGUMENT:

12.1 - Governments should do so only if such an action would definitely result in public advantage.

No. 13 - *A subtle system of ST classification should be elaborated on an international level, in order to reduce the user's time spent on searching.*

ARGUMENT:

13.1 - This would become another tower of Babel!

No. 14 - *Current awareness of ST information systems should be conceived in a way that automatically eliminates obsolete information in order to avoid unnecessary searching.*

ARGUMENTS:

14.1 - How to tackle this problem? One should publish less. Publishers should be more critical before they start to print. Some publishers sell their product with profit not because of its informative content but because of the publicity which they reprint!

14.2 - In the majority of cases there are no data bases which give an appreciation on the value of their information input. Trivial information coexists with valuable information. This leads to heavy and costly information pollution. (see also Argument 3.2 of Issue No.3.)

No. 15 - *A new "manual of international publishing" should be prepared. It should include the methods which have to be adopted when new publishing technologies are used. Supplementary or revised versions should be published each time when further new technologies have been implemented. The latest technological advances could be used for the compilation of such a manual.*

ARGUMENT:

15.1 - This suggestion presumes that the new technologies will completely supersede the traditional ones. It probably reflects the opinion of the producers of the respective equipment. This is why only time will show whether such a manual would be worthwhile.

No. 16 - *Technical standards, unified formats, standards of manuscript presentation etc. should be elaborated on an international level, so that the user will be able to take advantage of the new publishing technologies for worldwide communication and the integration of users and publishers from developing countries into larger networks and systems will be facilitated. Negotiations to establish such standards could be facilitated through teleconferencing.*

ARGUMENTS:

16.1 - Publishers and producers should care for such standardizations in their own interest. For the moment there are standards like ISO. Such methods can be followed and further adapted to new requirements. One should avoid standardizing, however, before the systems involved are fully developed.

16.2 - No, standardization should take place as soon as possible before too many different standards have had time to develop.

16.3 - Perhaps it would be easier to prepare and sell preprinted stationery which would include format instructions. This would be more effective than printed manuals.

16.4 - This is particularly necessary in order to enable "worldwide communication".

16.5 - The concept of "technological modifications", the basic issue of this symposium, should be better specified. We have not only to face modification of communication instruments and supports, but also a modification of *writing*. In the same way as when we passed from ideograph (hieroglyph/cuneiform) to alphabetic writing, we are now passing from alphabetic to electronic writing. Alphabetic writing will not immediately disappear, but a new form of writing will develop in parallel (telephotographical presentations, telephone, telematics).

Therefore, while discussing *standards* and *worldwide* communication, it becomes necessary to know whether alphabetic writing will still continue to be distributed according to the user's demand (or need). This is why a study on the "typology of written communications at the time of electronics" should be elaborated.

Consider that 80% of the potential "users" of any form of communication are illiterate. Are they to be educated by first learning the alphabet or directly through electronics ("alphabétisés ou électronisés")? This is why special attention must be given to countries outside the EC where electronics are not yet in use.

No. 17 - *Governments should encourage through economic measures the production of the technological equipment which corresponds most to the users' need, comfort and health. A government service could assume the function of a constantly open suggestion box and collect user suggestions and criticisms. Each time a new publishing technology has been implemented, this service could send out questionnaires to enquire into the state of acceptance by different user categories and transmit the information received to the interested parties to suggest possible improvements.*

ARGUMENTS:

17.1 - Governments are completely unfit for judging what is right or wrong. It is of course in the public interest to realize the most efficient information structures. This involves, however, being informed whether desirable methods are economically feasible. Governments can hardly judge this and would simply become victims of pressure groups.

17.2 - Users do not become users until the equipment is available to them. It is difficult for a potential user to theorize on his needs until he has explored potentials of the system. Therefore there is a need for a series of pilot studies with users designed to explore factors of reluctance, ease of use, needs in relation to task etc.

ISSUE No. 5:

THE INDIVIDUAL PUBLISHER'S FIRST APPROACH TO NEW PUBLISHING TECHNOLOGIES:
- WHETHER OR NOT TO INVEST?
- NEW BUSINESS VENTURES?
- HOW COULD GOVERNMENT INSTITUTIONS HELP?

POSITIONS

WHETHER OR NOT TO INVEST?

No. 1 - *The publisher's decision whether he should introduce new publishing technologies depends on the kind of information he is publishing. "Media should match matter", and the wrong technology for a particular kind of publishing could be counter-productive. First the needs of communicators and users must be clearly identified and then only the appropriate publication technology adopted.*

No. 2 - *Some new publishing technologies will be so costly in hardware and software investment and the hiring of adequate personnel that only major publishers can afford their introduction. Other new technologies will be sufficiently low in cost so that they will encourage the growth of small scale publishing. Existing information retrieval systems should be conceived so as to cope efficiently with these diversified developments.*

ARGUMENT:

2.1 - The basic question for the "traditional" publisher is that new technologies change the landscape in which he operates, and destroy familiar "monopoly" methods of operation that are the foundation of his business, e.g. classified advertisements *could* be diverted *from* local newspapers *to* viewdata: learned articles *from* journals *to* on-line distribution etc. The Financial Times, for example, is now in active discussion with computer specialists, TV set makers, 'chip' makers, disc makers, electronic storage specialists etc. *far away* from its traditional areas. Therefore, *no* publisher can ignore this "shift of the ground" under his feet: his 'product mix' must change. The great danger (as with photocomposition and electronic editing before it) is that everyone will "re-invent the wheel" i.e. make all the same mistakes: hence the role (of EEC, trade associations) to educate, elucidate.

No. 3 - *Some publishers will have to adopt new publishing technologies in order to remain competitive. If their financial situation does not allow the introduction of these technologies they could make arrangements with other publishers of their size in order to create common processing centres (EPCs) to share the investment.*

ARGUMENT:

3.1 - The nub of the publisher's problem is *not* that he can prove whether sales of printed products fall or increase: rather, it is that *other people* (e.g. on-line operators) are making money, in a pure commercial sense, from the publisher's material. Copyright is no protection against this. Therefore, if there is a demonstrable "market place" out there in the on-line field, normal commercial pressures (maximizing your income from your efforts) draw the publisher into this field.

No. 4 - *There is a danger that the small but efficient publishers who have something to communicate might be rationalized away. They should be protected, so that the patrimony of knowledge to which they contribute will be maintained.*

ARGUMENT:

4.1 - The small publishers can protect themselves from such a rationalization by furnishing those products which big publishers are not able to or do not want to produce, because they would lose their elasticity. Subsidies would lead to public waste and intellectual indolence.

No. 5 - *The publisher who decides to remain traditional will have to specialize. Within the continuous explosion of the information market, he will be able to explore wide and new fields of demand and maintain his income.*

ARGUMENT:

5.1 - How? Observe that his output can so easily be copied and resold without compensatory payments.

No. 6 - *The publisher who decides to move towards electronic production and distribution will have to consider in his investment calculations that he will have to maintain his conventionally trained staff to continue producing his traditional printed versions up to the point where the electronic media have found a secure market and push the printed products off the market. Then he may decide whether he will limit his production to the electronic product only or whether he will at the same time offer his traditional alternative to the electronically deprived areas in the world.*

ARGUMENT:

6.1 - The investment is strictly related to the publisher's financial capacity and to the nature of operation of his enterprise. The traditional way of publishing where a benevolent scientific editor assists the publisher may be more efficient than absolute modernism based on salaried staff.

6.2 - See Argument 16.5 of Issue No. 4.

No. 7 - *The publisher who introduces new publishing technologies must not risk buying hardware and software which look useful at first sight but will be obsolete tomorrow. A system which provides, for example, good print but bad image reproduction will not remain on the market for long.*

ARGUMENT:

7.1 - Here is a very wide gap. Very few publishers have enough expertise or reliable *unbiassed* advice. It is still largely a "hit or miss" situation. (See also Argument 12.1 of Issue No. 3.)

NEW BUSINESS VENTURES:

No. 9 - *The adoption of new publishing technologies permits the publication of material which would not be viable in print and can create a new market without diminishing the conventional market.*

ARGUMENT:

9.1 - Only experience could confirm this. "Depository" or "on demand publication" via microform would be such a possibility.

No. 10 - *Publishers who decide to move towards the electronic media could produce educational material for those who are unable to read conventionally printed media (spasmodically paralysed children, etc.).*

ARGUMENTS:

10.1 - Socially important minority group. Societies for blind could have more useful suggestions than publishers.

10.2 - Example: A computer transcription system into integral and abridged braille (TOBIA) is being developed at the University of Toulouse. It substitutes the mechanical information input by electronic input. Speed of output: 70 pages per minute, copies can be reproduced in unlimited numbers. Text quality is guaranteed through double control (input by a seeing person, output by a blind person). A publishing house for braille publishing will have to be created.

No. 11 - *The new technology publisher could stimulate the market for conventionally printed media rather than reduce it (publication of electronic catalogues on conventional books, etc.)*

ARGUMENT:

11.1 - Yes, provided the printed media have sufficient legal protection.

No. 12 - *Secondary on-line services could create data bases giving information on the publishing programme of traditional publishers. In exchange, the traditional publisher could stimulate the edition of monographs which show the utility of secondary services and new technology publishers to their conventional clients.*

ARGUMENT:

12.1 - This sounds somehow complicated and is certainly not realizable in such a form.

No. 13 - *The publisher who adopts new publication technologies should consider immediately the development of new by-products and their possible market (compile books from many sources, market reports, updatable handbooks, subsets of data banks, serving information seekers through electronic systems, offering electronic browsing facilities). He will be able to charge for information which is freely available elsewhere, because his clients will pay for using the electronic system.*

No. 14 - *Full text storage and distribution of business data could become a new industry of "pure" electronic publishing.*

ARGUMENT:

14.1 - Business data is an easily perishable merchandise because it quickly becomes obsolete. Maybe this is why electronic publishing would have a good chance because it offers quick access and enables quick data exchange and substitution.

No. 15 - *Publishers who have introduced new publication technologies with success could consider the extension of their experiences to the developing countries (hardware and expertise) which could become "information-starved" because up to now STI demand and supply emanates mainly from the developed world.*

ARGUMENTS:

15.1 - Developing countries should not be forced to run before they can walk - it will not work.

15.2 - This position sounds just as if the developing countries had identical and quantitatively just as interesting information needs as the highly developed industrial countries with their total differentiation of all functions and broad strata of educated population. Last but not least would be the problem of the language barrier. Where quantitative needs require the transfer of new information techniques, this will certainly be realized. But it should not be a matter to be treated with missionary-like ardour and subventions from development funds!

15.3 - A problem of climate will have to be observed, particularly when transmission of new information methods to developing countries is suggested, because the required hardware is extremely sensitive to temperature changes and will not be of practical use everywhere.

15.4 - Please refer to argument 16.4 of Issue 4 (80% of world population are illiterate - do we want to transfer first the alphabet or immediately electronics?).

HOW COULD GOVERNMENT INSTITUTIONS HELP?

No. 16 - *Governments should develop a coherent information policy and organize a good structure for regular consultation and cooperation with publishers.*

ARGUMENT:

16.1 - Governments should in fact do so. But injunctions should only be incurred in areas where commercial solutions would in no way permit the coverage of a clear public need. Example: providing electronic secondary services for information exchange, which could not be hoped to be cost-covering unless under exceptional conditions.

No. 17 - *A government service could collect information centrally on new technologies and disseminate it regularly to publishers and other interested parties. This service could also visit the individual publishers, study their present situation, and suggest solutions to their individual problems, based on technical, financial, legal and social facts.*

ARGUMENT:

17.1 - This is not a government affair - or does one suggest here to create new jobs?

No. 18 - *A government service could collect information centrally on new technologies from the different hardware and software producers and simultaneously send out questionnaires to publishers who have implemented these technologies in order to evaluate their practical experience and to transmit recommendations for improvement to the producers.*

ARGUMENT:

18.1 - This is not a government matter. It concerns rather the pertinent professional associations.

No. 19 - *A government service could establish the necessary connections between publishers (innovators and traditional ones), secondary services, libraries, on-line data bases, etc. to provide possibilities for cooperation instead of competition through division of tasks, mutual announcement of services etc. and to consider possible necessary changes of policies and legislation.*

ARGUMENT:

19.1 - This is not a government matter. Competition fulfils a very important function and should not be eliminated, otherwise it will be to the taxpayers' account some day. Government should not relieve entrepreneurs from decision-taking, if the problems concerned can or should be resolved through free economy. Only where this is not possible should a public solution be provided.

ISSUE No. 6:

FIRST ORGANIZATIONAL IMPACT OF ADOPTED NEW PUBLISHING TECHNOLOGIES ON THE PUBLISHER, THE PRINTER, THE COMPOSITOR, THE EDITOR AND THE AUTHOR, AND THE MODIFICATION OF THEIR RELATIVE ROLES.

POSITIONS

THE PUBLISHER

No. 1 - *The publisher's new equipment which may be computer and microprocessor-based includes: terminals, direct entry photosetters, electronic cameras able to produce halftones, word processors, both stand-alone and communicating, in the near future even voice input, OCR-equipment permitting simultaneous text and image processing and transmission on one single film to the printer.*

It will bring the initial production stages much closer together and permit: Direct entry by the author - Teletransmission to the editor/ publisher with direct printout of the text - Refereeing and editing including text corrections through direct tele-dialogue over terminals between publisher/editor/author - On-line transmission of final text to the printer.

ARGUMENTS:

1.1 - One too often forgets that the publishers' role is also an artistic one, that the lay-out is an art, and that the artistic quality of the product can have a great influence on the market, as shown in the evolution of school manuals. A totally programmed (automated) production system can hardly comply with this requirement.The data bases presently available are unable to handle big quantities of texts and above all make it difficult to search for anything that does not correspond exactly to the criteria established by the data base. As far as graphic or iconographic data is concerned, the case is completely hopeless.

1.2 - I agree with this position, but the statement is limited because it assumes continuation of production by traditional paper-based methods.

No. 2 - *The publisher is able through his new equipment to save a number of intermediary working processes and he can concentrate his efforts instead on increasing his choice of authors, editors and referees and open new fields to his publishing programme.*

ARGUMENTS:

2.1 - As hitherto, the publisher will positively not dispose of the producing firm simultaneously. This is why he will have to find the most qualified partner. But if he wants to produce by himself, the new technologies will first cause him supplementary burdens.

2.2 - Correct, but many people are working for publishers at home and most of the equipment at present in use is too bulky and too expensive to present a practical proposition.

No. 3 - *The publisher will be able through his new electronic equipment to create his own data base, establish at low cost constantly updated publishing programmes for different categories of user groups, provide output in any chosen form (microfiche, book, telecopy to data terminals, etc.) and satisfy on-demand requests.*

ARGUMENTS:

3.1 - This is a major opportunity, particularly for the smaller publisher.

3.2 - It depends on the publisher's programme. Such a realization would be worthwhile if it was not implemented just like a costly gadget but corresponded to quantitative needs which can be satisfied *only* through new or better technologies.

No. 4 - *To overcome problems of hardware and software compatibility between publishers, authors, editors and printers, publishers in general ought to cooperate in order to set up standards and possibly establish central bureau facilities for recording manuscripts.*

ARGUMENTS:

4.1 - This can be considered essential for the successful introduction of new information technologies and may require public (state) support.

4.2 - *Authors* should not be bothered at all with such problems. *Publishers, editors* and *printers* already have well functioning relations under the traditional techniques. Should standards be necessary, the interested partners will have to cooperate and oblige the producers of equipment to discontinue the creation of "in-house" standards.

4.3 - One or several such bureaus could be very helpful. Difficulties: lack of competent operating personnel, equipment costs.

4.4 - Central bureau facilities:

The suggestion made by one delegate that scientific publishing should be left to scientific societies is ridiculous and damaging. The reason why commercial publishers entered the market is because the institutional publishers were (and still are) so bad - long publication times, very nationally based, poor editorial development policies and little marketing ability. Concentration of scientific information dissemination in scientific societies would be disastrous for the scientific community, it would lead to monopolies and greater piracy of information.

4.5 - Central systems like the suggested one (Lockheed, for example) will not lead to monopolies of information. There are for example data bases like "Chemical Abstracts" which can be accessed at present through IRS/ESA, SDC, Lockheed, BRS, soon Infoline, DIMDI, Télésystèmes etc. No one system is able to manipulate the information of these bases, because any customary user would perceive it when comparing the results obtained over different systems.

No. 5 - *To overcome problems of personnel restructuration, modified author contracts etc., publishers will have to convene on new statutes which can cope with the situation created through the advent of new publishing technologies.*

THE PRINTER AND THE COMPOSITOR

No. 6 - *The records produced through the publisher's new input equipment are able to drive a variety of output devices, like:*

Daisy wheel printers, line printers, laser printers, phototypesetters, COM, intelligent copiers, ink jet printers, etc. Masters can be created for copying machines, small offset litho presses, etc., allowing the automated handling of masters and gathering operations;

which will cause important re-organization and job restructuration in the printing sector in general.

No. 7 - *As text-handling will become more closely associated with the originator and the publisher, a part of the printing operation will completely move away from the printer.*

No. 8 - *The function of the compositor will be taken over to a varying degree by the author or the editor.*

ARGUMENTS:

8.1 - Not until terminals are produced that can be used by authors at home.

8.2 - I think that two groups of *authors* have to be distinguished: the "hobby" author and the professional author (journalist). Nothing will change through the new technologies as far as the first type of author is concerned, except that he will possibly have his work published more quickly. This is why the *typesetter's* or the *compositor's* work will not become superfluous; it will change only in so far as he will have to provide the necessary control tags and transmit it to an electronic storage medium instead of performing the traditional typesetting.

It will be different for the second type of author (journalist). To him it makes sense if he types his article directly onto the data medium and assumes thus the typesetter's work (a second text transcription is thus avoided). But it is open to question whether the journalist should himself provide the text with software-dependent control tags or whether this work should be performed by a specialist.

An instruction table for writing manuscripts, standardized as far as possible, could possibly be provided for the scientific author too (first type) if he has the opportunity to type his manuscript directly on a data medium. The author could thus avoid errors which might occur when the text is copied several times by non-scientific staff (this is particularly important when non-latin alphabets or mathematical texts have to be incorporated).

8.3 - The function of the COMPOSITOR must remain. Modern typesetting methods demand a more complicated text composition than the traditional typesetting. This function requires not only a perfect knowledge of printing processes but also an exact knowledge of the computer programmes which transcribe the text into codes for the typesetting machines. It is particularly difficult to consider formatting aspects: the text has to be marked with control tags, because on the terminal screen it will appear in a different form than later when printed. In any case, the author would be overstrained if he had to assume such a function which requires special knowledge.

8.4 - The final text pages cannot be formatted on the terminal screen, because the screens presently available are too small (only 80 positions per line). The final presentation of the text impression can not be simulated on the display screen, because there all characters require the same space. Furthermore, only one single character size is possible. Tables and the like can only in rare cases be represented in a realistic form. It is difficult to create a good layout on the screen because one can hardly imagine what the printed text will look like afterwards (different character sizes, special signs etc., present difficulties).

No. 9 - *Printers and publishers will have to convene on standards, statutes and protective measures, in order to make the best possible use of the new publishing technologies.*

THE EDITOR

No. 10 - *The computer undertakes the basic technical editing. Editorial/ design functions merge because one operator is able to construct pages on a screen. Sub-editing can be accomplished on-line. Mechanical editorial jobs disappear, the others become more judgement orientated.*

ARGUMENT:

10.1 - The *editor* is part of the publishing function and occupies a far more creative role than appears to be appreciated: selection, improvement, presentation, accuracy, credibility.

No. 11 - *The editor of a scientific periodical (e.g. where the author is neither editor nor journalist, i.e. not a professional writer and not close to the final printed pro-*

duct) has to establish the prestige of his journal and will have to be helped by a number of scientists as referees to review the AUTHOR's text. The use of new publishing technologies (direct entry by AUTHOR through word processing) will rather complicate this procedure.

ARGUMENTS:

11.1 - Modern technologies, on the contrary, will facilitate this procedure because they permit direct interaction between all involved in generating a higher quality journal.

11.2 - The use of new publication technologies (e.g. direct text input) will not complicate but accelerate and simplify the refereeing procedure.

No. 12 - *Criteria for control during the editing process should be the following: a) Safeguard of quality rather than quantity and speed of scientific communications. b) Conversion to machine compatible form should not occur before the text has been refereed and recommended by the associate editor and accepted by the acceptance editor. c) Authority to change, update or modify stored text should be vested in the chief editor.*

ARGUMENT:

12.1 - The idea that the author should not be bothered with machine and software dependent data seems very important to me. Neither can one ask the author to change his manuscript each time depending on the publisher to whom he has submitted it. Therefore, the editor's role remains particularly important.

No. 13 - *Access to the stored information is of primary importance to the editor. If possible, each editor should have his own terminal on his desk.*

ARGUMENT:

13.1 - It depends on the kind of information and on the editor's requirements. Technical complications must not cost him supplementary time.

THE AUTHOR

No. 14 - *Word processing allows the author or inventor to change his text at any moment when he becomes aware of new findings. As the final text will be stored in the computer memory, without dissemination of paper copies, it can more easily be kept secret because only authorized persons will have access.*

ARGUMENTS:

14.1 - (This argument relates particularly to position 14 but concerns *all* positions of Issue No. 4.).

This issue mentions *authors*, *editors* and *publishers* in one breath. In my opinion, this is wrong. Editors and publishers are persons who are *pro-*

fessionally confronted with technologies of the publishing media. In the interest of the publishing houses and of the consumers they are bound to examine new technologies and, if need be, to implement them. The *author*, in contrast, is in general professionally a scientific person, an official, an artist, etc. but not a "*professional author*". Therefore, one cannot oblige authors who want to publish occasionally (sometimes only once in their life) to familiarize themselves with publication technologies. The author will continue to transmit his contribution on paper and will not care by which technologies his findings will be diffused. Editors and publishers as mediators between authors and readers will have the task of distributing these findings optimally and they have the professional task of applying, if convenient, the new technologies.

14.2 - Keeping information secret with the help of word processing is a remote possibility which relates more to "non-information" than to "information".

14.3 - Not every author wishes to familiarize himself directly with the new technologies. A university professor, for example, who is used to handing his manuscript over to his secretary to type a fair copy to be transmitted to the publisher, will not wish to make any modifications on a terminal himself. Therefore, the data input over terminals would have to be done by the secretary, but this will cause employment problems if too much specialized knowledge on the text processing programme and the operation of control tags was required.

No. 15 - *Decentralized word processing opens optimal possibilities for team work by several authors. The members of the team can easily communicate through a common data base about the progress of their work.*

No. 16 - *Decentralized word processing has a considerable impact on the organizational and financial structure of the production process. Multiple legal questions will have to be resolved, for example, how can the creative share of the individual members of the author team be determined?*

ARGUMENTS:

16.1 - If the information is not patentable, and the contribution of each author can easily be determined from the beginning, the authors will hardly be disposed to appear as a team without a clear prior agreement. It is in the line of human nature that each one considers his own creative part higher than the one of of his co-author's.

16.2 - I agree that the impact of a journal which consists of many articles written by many authors can be evaluated statistically by the number of times the journal titles are searched for over a data network. I strongly *oppose* and *contest* evaluation of individual research workers by this method. It stands to reason that the really original and creative papers may need a long lag period before they get an impact measurable by such a system - they may need evaluation by personal contacts and personal judgements and laboratory tests before they are "discovered" on a large scale. At the same time originality and creativity constitute the centre of the whole research system, but I do not think they are easily assessable as computer outputs.

No. 17 - *Authors will certainly make an effort to cope with the requirements of modern publishing technologies like direct manuscript input and adaptation to formats etc., because these technologies allow an almost immediate publication of their text and attract the attention of a much larger audience than the conventional one.*

ARGUMENTS:

17.1 - Maybe, but it is psychologically more acceptable for authors to see *printed* copies of their work.

17.2 - One must not neglect the human aspects of the communication process. There is a possible danger of an author inputting his material directly onto a machine - he may not go through the necessary precondition of exposing his material on a limited basis to colleagues who can help shape and develop his ideas (i.e. his own personal "refereeing" system).

17.3 - Viewdata systems may (especially if politically encouraged) enable certain authors and commentators (especially those whose material can be expressed in the narrow viewdata format, and needs changing frequently) to write directly for, and communicate directly with, his audience, under his own control and pricing, thus bypassing traditional "conglomerate" formats such as newspapers, journals etc. which are under the control of an editor. The economies of scale that apply to traditional printing, and dictate "conglomerate" forms of publishing, do not apply to viewdata systems. *A new opportunity* for authors.

No. 18 - *The author's reputation (and social status) is largely associated with his "visibility" in the scientific community. Some new technologies (e.g. microforms and on-demand-systems) tend to bury his visibility.*

ARGUMENTS:

18.1 - "Visibility" is essential for the author. New technologies will have to take this into account and must not set aside authors for information technological reasons.

18.2 - If an author is above average he will not be buried in such kinds of publication. To the other authors it is not of importance whether they are visible or not.

18.3 - During the Middle Ages and before, authorship (literature, cathedrals, etc.) often did not reach beyond the region and the period: we thus do not know the names of these poets, playwrights, artisans, etc. or are in doubt about them.

In more recent times we are witnessing - especially as regards information - a shift from authorship to subject matter. To my mind, present-day technology will intensify this shift, and make the author more anonymous still.

18.4 - Publication of primary scientific information in the printed form is vital to most academic authors since their reputation and career prospects depend not only on the quality and quantity of papers written but also on where they are published. New technologies cannot have the same appeal to authors simply because they have no graphical or physical form and no prestige.

No. 19 - *Proposed linkings of authors with more speed and efficiency to the publishers (CRT terminals and other new technologies) will generate a "caste" system among the authors. Young, unknown authors with no corresponding technical equipment will be confronted with more difficulties in becoming visible.*

ARGUMENTS:

19.1 - To avoid a "caste-system" among authors, it is necessary to maintain the former possibility of transmitting a typewritten manuscript to the publisher.

19.2 - Maybe this is a reason also why a direct linking with the author should be avoided. It is fairly improbable that the publisher decides to let the author transmit a text unknown to him directly onto the machine ready for print.

ISSUE No. 7:

REVIEW OF SOME NEW PUBLISHING TECHNOLOGIES WHICH ARE ALREADY USED FOR PROCESSING, PRODUCTION AND DATA STORAGE, THEIR ADVANTAGES AND DISADVANTAGES, AND SUGGESTIONS FOR IMPROVEMENT:

- (WORD PROCESSING - PHOTOTYPESETTING - MICROFORMS - HOLOGRAPHIC STORAGE - "WORKING FROM HOME" - LASER PRINTER)

POSITIONS

WORD PROCESSING

No. 1 - *A word processor provides constant availability of the latest text which can easily be edited and updated. It provides direct access and flexibility of reproduction (it can format, print, condense, justify margins and collate).*

No. 2 - *Some word processors are not designed with photocomposition in mind and expensive interface programmes must be prepared to bring out their potential value. Those designed for photocomposition are incompatible with all but their own machinery. Standards to avoid re-keyboarding are necessary, or at least, interface material ought to be standardized.*

PHOTOTYPESETTING

No. 3 - *Phototypesetting compared to manual typesetting is less expensive and time saving, because text input and corrections can be done in one single working phase. Assembling of indexes, evaluation of comparative data and scientific evaluations can be operated automatically. Previously stored data can be added directly to the typesetting programme.*

No. 4 - *In phototypesetting texts normally subject to errors when manual typesetting is used (listings, foreign language, non-latin alphabet) can be processed more reliably.*

No. 5 - *Phototypesetting offers additional exploitation possibilities, e.g. full text storage and subsequent production of cheap extra editions which can be sold to other users.*

ARGUMENT:

5.1 - Separate production of such texts will be expensive and demand a complicated organization.

No. 6 - *Publications produced through phototypesetting and promising several editions can be stored at low cost, corrections and modifications can easily be added. Additions to text are possible at shorter intervals and the information can be kept updated more easily.*

ARGUMENT:

6.1 - This applies particularly to telephone books and similar works, like reference books etc. which have to be reprinted again and again.

No. 7 - *The control and command of the phototypesetting device requires an exhaustive and sophisticated programme. Text will need to be marked with numerous command tags for the typesetting programme. Therefore, thoroughly trained and experienced personnel will be necessary.*

No. 8 - *In phototypesetting much experience is required in order to recognize and introduce the aesthetic aspects at the moment of text input, another necessity for qualified personnel.*

ARGUMENT:

8.1 - This is why for many kinds of applications phototypesetting will not be worthwhile.

No. 9 - *For the coupling of word processing and phototypesetting, standards have to be elaborated, particularly for compatible data transmission units (floppy discs).*

MICROFORMS

No. 10 - *Microfilming is cheap compared to traditional printing. Micro-computers are getting cheaper. Complete text storage is possible. Very limited storage space is required. It provides a perfect basis for on-demand publishing.*

No. 11 - *The present use of microform could be lifted to a much greater height through:*

- Improved technology in the production of readers and reader/printers and consequently of picture copyflow. Improved quality of film/fiche.

- Setting up of standards for presentation on microform, especially fiche.

ARGUMENT:

11.1 - This is a field where producers of equipment are to be blamed most. They do not want to standardize in order to commit their customers to themselves. This stupid behaviour has been going on for 35 years already.

HOLOGRAPHIC STORAGE

No. 12 - *Holographic storage is more durable than magnetic tape storage; it eliminates the need for constant updating of magnetic tapes. Its capacity permits a greater density of information storage.*

ARGUMENT:

12.1 - Note that "holographic storage" is not another storage method, but only a specialized version of optical storage of machine-readable information. For technical reasons its application may not mature as fast as other versions of optical storage which may have a direct impact on STI (e.g. digital optical disc).

No. 13 - *Holographic storage is still an untried technology for STI. Access to data will be more complicated than to magnetic tape.*

No. 14 - *It is possible that some of the future growth in magnetic tape data base publishing might be "creamed off" by the development of holographic systems.*

DEVICES FOR "WORKING FROM HOME"

No. 15 - *Working from home: develop cheap keyboards with limited (microtape) storage using home TV screen.*

ARGUMENTS:

15.1 - Development work is needed to enable an author to prepare his material *at home* on a compact portable terminal either producing a magnetic record on the spot or connected via the telephone line to the producer.

15.2 - Working from home has two aspects:

a) Generation and storage of work in machine-readable form with inexpensive devices and use of standard interchangeable storage media (e.g. floppy disc). or

b) On-line working based on low communication equipment and time cost. For reasons of independence and economics (a) appears more desirable.

No. 16 - *Working from home: develop "interactive briefcase" for less than £1,000.*

ARGUMENTS:

16.1 - Examples of present research: (extracted from file transmitted by P. Cornelius)

- Texas Instruments TI 765 (20 to 80 KB);
- German Patent Application No. 27, 39, 157, dated 31.8.77, published on 15.3.79 (inventors: FAERBER/MARKO), (20 to 80 KB);
- SIEMENS-ROCKWELL cooperation for development of 256 KB storage elements;
- INTEL Megabit storage modules;
- IBM/BELL research.

16.2 - See also position 24 of Issue No. 6.

LASER PRINTER (e.g. IBM 6670 and similar devices)

No. 17 - *The high speed laser printer (e.g. IBM 6670) coupled with a built-in copying device can produce a clear original on plain paper and on both sides of a sheet.*

ARGUMENT:

17/18.1 - If the development of a laser printer is really worthwhile it will appear on the market, be it IBM or not.

No. 18 - *A new typographic design for the laser printer (e.g. IBM 6670 and similar devices) is suggested, i.e.:*

- *Adaptation to well established practices of typographic layout including readability research.*
- *Development of suitable fonts and character sets.*
- *Development of algorithm for context-sensitive typographic page make-up.*
- *Methods of fast storage and retrieval of lay-out formatted text to allow set printing.*

ARGUMENT:

18.1 - The EEC should encourage development and marketing of a European hybrid OCR/copying device which may:

- act as digital facsimile transceiver;
- switch to OCR for recognizable formats of type;
- serve as output device, with multiple formats and point sizes specified in software, for word processing systems;
- be used as conventional copier;
- handle collation + double-sided printing automatically;
- work over PSTN at line speeds of at least 9 600 wps, upgrading to 48 000 wps.

ISSUE No. 8:

REVIEW OF NEW FORMS OF OUTPUT, COMMUNICATION AND DISSEMINATION MEDIA, THEIR ADVANTAGES, DISADVANTAGES AND SUGGESTIONS FOR IMPROVEMENT OF TECHNOLOGIES, METHODS AND POLICIES.

(MICROFORM - VIDEOSYSTEMS - TELECOMMUNICATION - THE ELECTRONIC BOOK)

POSITIONS

MICROFORM

No. 1 - *Microform allows the user to receive at his place of work the maximum of regularly required documents, permits regular substitution of obsolete information and constitutes a perfect basis for on-demand publishing.*

ARGUMENTS:

1.1 - The user will probably prefer a minimum to a maximum of information, i.e., the really required amount of information, but nothing superfluous. This is why it is not the information medium which forms the basis for "on-demand publishing" but the author's qualification and his judgement.

1.2 - Yes, if "publishing" is seen as "information service", providing up-to-date personalized library maintenance - organizationally and technically feasible by integrated use of microform and data base publishing and if the user can be re-educated to pay for the "information service", not simply for "documents".

No. 2 - *Microform does not allow browsing.*

ARGUMENT:

2.1 - Microfilm *does* allow browsing. May be even better and quicker, especially for newspapers, on an automatic reader.

No. 3 - *Microform: Technologies of readers and reader/printers should be improved.*

ARGUMENT:

3.1 - Small portable microfiche readers (size and weight of a portable typewriter) have been on the market for several years.

Reluctance to use microforms very often originates in the fact that users ignore the quite simple way of reading microfiches with the new portable readers, even at home.

No. 4 - *In general, microform versions of ST literature are disseminated with too much delay. Publishers should abandon their practice of withholding the microform version for months and years after the issue of the print version.*

VIDEOSYSTEMS

No. 5 - *Videotex will facilitate wider accessibility of information than traditional publishing, because:*

- *"Specific" information can be accessed more cheaply.*
- *The very latest information is provided and can be constantly updated.*
- *It can interconnect computers and local, personal, computing and processing facilities and can become an access point for further information searching.*
- *It can give access to microform libraries.*

ARGUMENT:

5.1 - This statement is somehow unrealistic. I do not mean that such services could not be technically realized, but I doubt that it can be done in *such* a way. After all, videotex presupposes a certain *mass* utilization.

No. 6 - *Videotex will first encroach upon and later supersede some traditional forms of publishing.*

ARGUMENT:

6.1 - In response to a plea for the maintenance of printed books in the face of the growing proliferation of non-book publications (microfilm, videotex, electronic books, etc.):

So far as concerns intergovernmental bodies such as the World Health Organization, which have to publish to meet the demands of a wide range of cultures and readers the world over, in more than 150 Member States, it is most unlikely that the printed page will be superseded for many years, although the new technologies may well be used for the preparation of that page. The provincial health administrator, the teacher of rural midwives and her pupils will continue to require material that can be read by the unaided eye, without electricity intervening. at least until the "non-book" has universally replaced the present true book.

This does not mean that new technology outputs will not be used, indeed they are already used for some purposes; but rather that the book will remain the basic means of communication, supplemented by others for special purposes.

No. 7 - *Videodisc technology takes advantage of technical developments in other electrical/electronic consumer industries. It gives possibilities for interfacing STI videodiscs with other new visual information systems (videotex, on-line systems).*

ARGUMENT:

7.1 - (Refers also to position No. 7).

The statements are not true, or only with limitations, for videodiscs as planned for the consumer market. The statements become fully true for the next generation, the optical disc, which will be derived from the digital audiodisc (ultra-hifi-record) to be introduced during the early 80's.

No. 8 - *Videodiscs can be produced (stamped) at low cost. This may overcome the problems of updating which are inherent in the disc publishing industry.*

No. 9 - *Videodiscs for STI are being actively researched into in the U.S. The results of these investigations might revolutionize primary publishing.*

No. 10 - *Viewdata and TV displays have a very limited capacity compared to print media or microforms. This will not limit data retrieval but it constitutes an important limitation for study type use (continuous reading). Therefore the use of videosystems is not recommended for scholarly publishing.*

ARGUMENT:

10.1 - Yes, but a modification of systems approach in the context of personal computers may reduce some limitations.

No. 11 - *The problems caused by the limited frame capacity of viewdata displays can be overcome through reformatting.*

No. 12 - *High resolution displays with high character capacities should be developed to make viewdata usable also for scholarly publishing. A feasibility and cost study of such a realization is recommended.*

ARGUMENT:

12.1 - You cannot browse, you cannot collate - all this can be held against such a suggestion.

No. 13 - *Data base on-line systems could be linked with viewdata systems and a mass market for certain types of STI could be created under the condition of:*

- achieving compatibility between analogue and digital systems;

- reconciling the pattern of "menupages" (viewdata) and sophisticated indexing/thesaurus (data bases).

ADVANCED TELECOMMUNICATION SYSTEMS

No. 14 - *STI transfer takes place through "informal" channels (correspondence, reprints, conferences). Teleconferencing through constantly better and cheaper telecommunication facilities will reduce the costs of conferencing, stimulating thus the informal methods of ST communication, but it will decrease the publisher's market for "formal" STI information.*

No. 15 - *Teleconferencing could give the Third World an opportunity for equal access to "Centres of Excellence".*

ARGUMENT:

15.1 - More so as an auditor than a participant. And would this be so much cheaper than the present ways of information access offered to the Third World?

No. 16 - *Adoption of fibre optics by the PTTs should provide more flexibility for message carrying.*

No. 17 - *Improved telecommunication facilities will permit the introduction and coupling into networks of full text data bases with overall access (experiments underway in the U.S.).*

No. 18 - *Digital fax transfer through telecommunication networks will open unlimited distribution possibilities and necessitates a strict regulatory framework to control copyright infringements.*

ARGUMENT:

18.1 - Only if wideband telecommunication becomes economically accessible to many.

THE ELECTRONIC BOOK

No. 19 - *Portability of non-book literature still poses the problem that some items (like tapes) need presentation devices and that most of them are not standardized and unlikely to be covered through ISO rules.*

No. 20 - *Non-book educational literature poses the problem that educational programmes for micro-computers and even for disc calculators cannot be transferred from one school to the other.*

ARGUMENT:

20.1 - Within a few years home computers will be very widespread and accepted as part of everyday life. With them will come an explosion of self-teaching modules covering everything from knitting to nuclear physics. This opens up both opportunities and threats to the publishing trade:

- *Opportunities* in the vast new market of people eager to learn. Software on videodisc or tape will allow subjects to be taught that would not have been tackled with book reading. Opportunities for publishing texts to accompany software modules.

- *Threats* to those publishers currently in the education fields, as many types of books will not sell in competition with audiovisual software (on both price and desirability).

No. 21 - *Digital text storage allows the printout of just the required chapters of a text, either locally or remotely over the telephone network and is particularly suited for tailor-made information distribution.*

No. 22 - *The flat liquid crystal screen could increase the readability of non-book formats. It could enable a device of book size to read and display from diskette or cassette.*

No. 23 - *The briefcase computer represents an easily portable electronic book device. Stored information will be transmitted with high speed on-line from full text data bases into the magnet bubble storage of the briefcase computer equipped with a liquid crystal display screen and supplied either by the mains or a battery. The text appears on the screen or will be printed out on demand.*

No. 24 - *The briefcase computer can furnish literature for the home library in the following ways:*

- *COM offprint or hardcopy printouts furnished by the data base;*
- *connection to the home computer and recording of the information onto a personal long duration storage medium.*

ARGUMENT:

24.1 - (Concerns positions No. 24 and 25)

. *Objection*: relatively expensive addition (at present) to a terminal of the Prestel/Télétel type.

. *Alternative*: quick and cheap photocopy from screen. "Prestel" representatives say this is under examination (but probably only for black and white).

. *Recommendation*: should be encouragea but (again) raises copyright/royalty questions.

No. 25 - *Silicon chips will be used to store the contents of whole books at a very low cost. The chips will be plugged into a high quality display device of traditional book size.*

No. 26 - *The silicon chip book needs a minimum of space. Whole libraries will occupy the space required by one paperback. Distribution of the silicon chip book will be through dispatch of chips or through telecommunication recording onto chips. It will cost approximately 1/50 of a traditional book.*

No. 27 - *Advances in silicon chip technology will produce "smart" books that will do their own research and presentation of ideas.*

ARGUMENT:

27.1 - The chip may be cheap but the software is not.

ISSUE No. 9:

JOURNALS AND NEWSPAPERS CONFRONTED WITH ELECTRONIC MEDIA.

HOW CAN GOVERNMENT INSTITUTIONS ENSURE A GOOD OVERALL ACCESS TO ELECTRONIC PUBLISHING?

POSITIONS

THE ELECTRONIC JOURNAL

No. 1 - *Journals with a large circulation will not essentially be touched by the restructurations of the information scenery and will not need to move towards electronic distribution.*

No. 2 - *The electronic journal is most suited to "scientific and technical" literature and data, either to which very fast access is required or of which very little use is expected (e.g. a particular and restricted circle of subscribers requiring more tailor-made information). It is well suited for on-demand publishing, synopsis publishing, etc.*

No. 3 - *The electronic journal is not suited to most "non-scientific" literature nor necessarily for ST literature that is likely to receive substantial use (for which conventional publication is economic) or for which widespread visual exposure is desirable.*

ARGUMENT:

3.1 - (Concerns also positions No. 2 and 3).

It is difficult to see how any text system will understand the semantics of the text it is handling: it is not at all clear that scientific text is alone in being suitable for electronic handling.

No. 4 - *The electronic journal is psychologically and ergonomically ill-suited to general use as a medium of ST dissemination. Browsing is not feasible. Authors, referees and subscribers need a terminal. It cannot be read on a train, a plane, or in bed.*

ARGUMENT:

4.1 - Browsing is feasible. The computer, by 'learning' the types of information examined by a particular user in the past, can present to him new information which meets his past selection criteria - i.e. browsing. The electronic journal can be loaded onto a 'briefcase terminal' (see Positions 23 and 24 of Issue 8) and then be taken anywhere.

No. 5 - *Worldwide access to electronic ST journals will be difficult because hard and software standards are lacking. How could a reader from a Western country get access to an electronic journal issued in an East block country? How could it work the other way around?*

No. 6 - *The electronic journal is cheaper to produce than the conventional print-on-paper journal. Input and access is very rapid. It provides a facility for detailed indexing, classification, etc.*

No. 7 - *The electronic journal will present storage problems with regard to its back-up files over a number of years.*

ARGUMENT:

7.1 - The New York Times inquiry service manages very well, handling more than 80% of all requests, with only approximately 2 years of files constantly available. Laser holographic storage will make a nonsense of this statement.

No. 8 - *The electronic journal requires a network different from the ordinary library system. It will present severe problems as far as evaluation by abstracting services is concerned.*

THE ELECTRONIC NEWSPAPER AND THE NEWSLETTER

No. 9 - *Newspapers and magazines are social and political organs, as well as information channels. Computer databases are (or are so far) neutral and emotionless. Therefore, NEWSPAPER publishing should remain traditional.*

ARGUMENT:

9.1 - Is a non sequitur; TV news manages very well to be social and political.

No. 10 - *It is important that newspapers retain their social/political roles, whether traditionally printed or electronically published. Measures should be taken to safeguard this role.*

No. 11 - *Modern technologies tend to favour concentrations of newspapers, through need of concentrated investment funds, necessary standardization and therefore deeper hierarchical control. This could hamper the freedom of expression and independence of a newspaper.*

No. 12 - *Newsletters on the "issues of the day" could quickly and cheaply be produced and disseminated through electronic media (digital fax, etc.). Three hundred years ago, the correspondence between scientific adversaries used to be published and distributed - the same ought to be done today, by modern means.*

ARGUMENT:

12.1 - I hope not: one should realize after all that people want to dedicate only a limited amount of their day to information. We are being drowned anyhow by the bulk of information offered, and the beautiful idea exposed in this position presupposes the readiness of those possibly interested, in spending, not only money for such nice things but, worse, to sacrifice their time.

HOW CAN GOVERNMENT INSTITUTIONS ENSURE A GOOD OVERALL ACCESS TO ELECTRONIC PUBLISHING?

No. 13 - *Support projects to study in which form primary information is most suited for direct electronic transmission (e.g. in synoptic form?).*

ARGUMENTS:

13.1 - A detailed cost-benefit study should be established in order to find out *which* of the publication technologies (paper print, microfiche, machine-readable storage media like magnetic tape or disc, holographic storage, video discs, silicon chips etc.) are economically the most advantageous ones for the different areas of applications in the publication sector.

13.2 - Governments should support such a study only if it was in the public interest. Is this so? Are the presently existing media (including TV and broadcasting together with the printed press) not sufficient already?

No. 14 - *Support technological projects on:*

- linking a large number of separate electronic systems in one cohesive network;

- developing efficient and inexpensive international networks without the need for "digital-to-analogue" and "analogue-to-digital" conversion;

- development of management information systems for monitoring a worldwide network.

ARGUMENTS:

14.1 - The best way the EEC can ensure overall access is by developing and insisting on rigid adherence (probably in conjunction with ISO + CCITJ) to standards for information transmission, line protocols and message formats.

14.2 - Would this really be in the users' interest? Or is it only in the interest of those who want to develop interesting new possibilities?

No. 15 - *Financially support the creation of worldwide international network operations (covering both developed and developing countries).*

ARGUMENT:

15.1 - Why *worldwide* networks? There are already so many language problems.

No. 16 - *Promote compatibility in indexing vocabularies and query language.*

ARGUMENT:

16.1 - Yes, this should be done. Among others, UNESCO is working for these purposes.

No. 17 - *Stimulate the development of national and supranational information policies and the necessary compatibility between these; encourage policies promoting innovation and investments by the information industry.*

ARGUMENT:

17.1 - Innovations and investments in the information industry shall not be determined by governments but by the interests of the market. Otherwise the taxpayer will be badly served.

No. 18 - *The role of the state in providing much of the infrastructure of electronic publishing (e.g. Euronet, packet switching, viewdata computers and satellites) should not infringe the independence and at times partisan approach of the printed media. (The common carrier or "open house" position of the British Post Office on its "Prestel viewdata service" is in this context a model to be followed.)*

ARGUMENTS:

18.1 - I agree with this position but I cannot comment on this advance praise for the "Prestel" system.

18.2 - It should be discussed in a European context how the social science/education publishing world could best link with the developing teletex/viewdata systems.

ISSUE No. 10

HOW CAN LIBRARIES AND SECONDARY SERVICES USE THE NEW TECHNOLOGIES AND HOW CAN THEY COOPERATE WITH PUBLISHERS?

POSITIONS

THE LIBRARY

No. 1 - *Advantages for a library moving into new technologies: Through on-line connections, facsimile transmission etc., interest profiles can be more easily matched with deposited collections, tables of contents of newly appeared editions can be checked on a terminal or by telecopy (this facilitates the decision whether a new acquisition would be worthwhile or not). Subscriptions and copy delivery can be treated through teleordering. Storage space can be saved through microform, videodiscs, etc.*

ARGUMENT:

1.1 - As director of a large UK university library I wish to enter a plea on behalf of the *expressed needs of the 'consumer'* of information - in my own case the university student, research worker or teacher.

I see no immediate sign of an overwhelming demand for alternative modes of presentation of information, even in a university which is biased in favour of science and engineering. Indeed there is in many areas of university activity antipathy or hostility to non-traditional modes - even extending to the University Library's COM catalogues. Hostility is undoubtedly prevalent in the humanities and certain areas of social science where traditional texts are unlikely to be superseded.

I am aware of the arguments in favour of new technology. I am also conscious of the fact that economies of operation must be found - but these must be found not only by the information providers, but also by the users, and the high initial costs of launching the new technologies in a library must surely be justified by a perceived need for such benefits as they may bring.

No. 2 - *The present role of the specialized ST library as custodian of printed documents will diminish or even disappear. It will become a service orientated organization locating information and answering questions for its users, many of whom will not be specialists (therefore library staff will have to be highly qualified).*

This new kind of library will no longer be able to serve its clients free-of-charge. Therefore the user's free access even to hardcopy and microforms might become undermined, which poses an ethical problem.

ARGUMENTS:

2.1 - It appears very important to *adjust the education of documentalists and information scientists*, but not to let the profession of the librarian disappear. The librarians should know how to deal with the new terminals and how to access all the data bases in the different fields of interest of the library clients.

2.2 - There is a danger of overlooking the design or even the identification of the correct environment in which these new forms of technological transmission systems will come into play with their users and this will happen at different times in different parts of the world.

Environment is all important. To take the example of Euronet, it has been said that the presence of practising librarians as part of the design team was noticeably absent, yet it is now these librarians who are being expected *to present the service to potential users,* all too often in an environment ill-equipped and ill-designed to do it. But Euronet is only one example of the new technological revolution in information transfer - more sophisticated and unconventional transfer products will have even more far-reaching implications for the environmental designers.

2.3 - The right to information must not become restricted by commercial or financial barriers. Information is a part of the educational process, and equal access to it should be secured for everybody. Therefore the suggested kind of new library will not so much raise an ethical problem, but require fundamental decisions to be made as to what is the public interest.

No. 3 - *Disadvantages of new library systems: Hard copy documents can be acquired and stored permanently, but information received through interactive SERVICES could be lost due to system defects, labour disputes, political decisions and financial failure. Thus, the library that moves completely into new systems can no longer offer a secure service to the user unless all documents received through interactive services are immediately reproduced and stored in a durable way. This requires a sufficient number of reproduction and storage facilities which guarantee access for everybody at any time (e.g. a sufficient number of terminals and microprinters/readers). Consequently, space for sufficient technical equipment will have to be foreseen in library building design.*

ARGUMENTS:

3.1 - Perhaps it has become necessary to set up a special and representative working group to consider all aspects of the environmental problems which will still be with us after the technical production and transmission problems have been settled.

3.2 - Where is the sense of representing such "new libraries" and possibly promoting them, if their disadvantages are so important and their usability could be restricted by any fortuitous event, and if, as a whole, only new dependencies and inconveniences are created which will not serve but handicap the communication of knowledge.

3.3 - Together with the development of new modes of information presentation, research should continue into cheaper and more efficient methods of producing conventional text and making it available to the library user. Between the challenge of technologies and the opportunities of the market, the real needs of the user must not be lost to view. Librarians and publishers, being interdependent, must cooperate and try to achieve the same objectives in order to overcome the large gaps that exist today between the information provider, transmitter and user.

THE RELATION PUBLISHER/LIBRARY/SECONDARY SERVICES

No. 4 - *Secondary services are most suited to move away from printed form. Through new technologies, they can increase the quantity and quickness of their announcements and multiply their ability to supply a more and more refined selection of tailor-made information packages.*

ARGUMENTS:

4.1 - In this context, the *creation of secondary information services for the legal sector* is suggested and justified as follows: (Refers also to Position No. 10).

a) *Quantitative need*: there are hundreds of thousands of attorneys, courts, tax counsels, chartered accountants.

b) *Qualitative need:*

- In macro-economic terms, because of

. forthcoming free circulation of attorneys in the EC;
. increasing need of legal aid on a European level;
. increasing complexity of law in general within the EC.

- In micro-economic terms, because the traditional storage of legal information (volume of legal documents and text books) absorbs too much time, personnel and energy.

c) *Fields of application for on-line connections:*

- *Legal sources:*

. Texts of laws and regulations;
. Case law and model sentences.

- *Abstracts* (also in translation) of specialized journal articles.

- *Technical information* (lists of attorneys, courts and correspondent attorneys to facilitate search for expertise).

d) *Prospects for publishers:*

- Cooperation within a European pool.

- Smaller publishers who have the care of certain special fields (e.g. foodstuff laws, game laws) will also be able to participate.

- Parallelism of the secondary system, thus no competition with the printed products.

4.2 - *Suggestion for the creation of secondary services in the social sciences sector*: (Refers also to Position No. 10).

During the 1960s the growth of computerization of *secondary services* in the "hard sciences" (chemistry, physics, biology, medicine) has accelerated. A similar trend in the development of *secondaries* in the social sciences has been anticipated within a European context (Council of Europe); over the last decade, NFER, for example, has established a UK data base for *educational* research, computerized this, and in conjunction with the Council of Europe provided sub-sets of the information to *eudised* (hard copy outputs: "UK Register of Educational Research", "The *eudised* R & D Bulletin").

4.3 - *Use of videotex* for *secondary services* and *libraries:*

The new techniques, such as videotex, could be used to give public library readers a better say in book selection (via the national bibliography, which would have to be produced on viewdata). The reader, having made his selection, would dial the library's viewdata number and, citing the new books that he wished to read, say 'get this'. His desiderata would be taken into account when books were purchased for the library.

If this facility were taken up on a large scale, it might create a change in the pattern of demand, with consequent repercussions on publishing.

An advantage of this procedure is that it dispels the Orwellian atmosphere of '1984' which at present surrounds new electronic techniques, including viewdata; at least one process would be democratized, rather than the reverse.

No. 5 - *Through the use of new technologes, libraries and secondary services can more easily provide an on-demand supply of copies in any chosen form. Therefore, the publishers' sales will diminish. To maintain their income, publishers should be compensated through royalty payments for each delivered photocopy. Libraries consider such royalty payments as a limitation to the free access to ST information.*

5.1 - An increasing volume of library services in the future will be *on demand* (with enormous benefits over conventional library practice).*Libraries* can no longer defend access to free-of-royalty copy.
Some of the reasons are as follows:

- libraries already pay for interactive services;

- libraries in the UK are already required to recompense publishers for loans (the Public Lending Right issue);

- some organisations (the Copyright Clearance Centre in the USA and Anbar Abstracts in the UK) already recompense publishers for copying.

Most Western European law libraries are already required to charge an economic amount for photocopying - all that is required therefore is an additional component for royalties.

In the new licensing system libraries will forfeit the right to free-of-royalty copying, copyright owners having the right to withhold consent.

No new medium (including videodiscs which will be more subject to consumer copyright than institutional copyright) will escape the problem of copyright.

5.2 - To have immediate access to a journal or journals which publish papers in one's own field is very important and a *library is essential for this reason*. We in King's College are involved in research on the periodical literature in the physical sciences. This is a quantitative study, and surveys the last 25 years. It has as a base a core of journals, 150 titles in all, on which quantitative and statistical calculations are performed, and the impact of new technology assessed.
In some cases, however, the *opposition* on the part of commercial *publishers* to provide information regarding the journals published by them makes our efforts very difficult.

(See an abridged form of our preliminary results published in Nature, Vol. 279, 17 May 1979: Correspondence: "Is physics levelling off?".)

5.3 - Photocopying and the specialist journal: Those who claim that photocopying is killing the specialist journal should produce evidence for that statement. In spite of a number of studies which expected to show this is the case, so far most have come up with negative results. (See for instance Allen Kent and Bernard Fry's study of cancellations of journal subscriptions). This is not to say that people should not pay for the content of the photocopy as well as the paper, use of machine, electricity, etc.

5.4 - New technologies permit easy and cheap reproduction of documents. An important part of the activity of *secondary services* consists of furnishing the documents which they have announced. The same applies to *libraries* with respect to the literature they hold. The problems raised by this situation have been evoked: we have heard of publishers driven towards failure. *Publishers* wish law to be amended to include the payment of royalties on photocopying and reproduction costs. Being in opposition to such amendments, do *secondary services and libraries* really uphold free circulation of information? There is conflict to which new technologies have led by provoking a rupture of the existing commercial equilibrium between publishers and secondary services.

5.5 - Another possibility of paying royalties to the *publisher* is in transmitting to him a pro rata share of the remuneration received per bibliographic reference. This could be based on publishers transmitting machine-readable abstracts and bibliographic data of their publications to the *secondary services*. The bibliographic data bases would thus receive a part of the primary information (i.e. abstracts which would have to be remunerated accordingly).

5.6 - There has been a shift in library text-material from the scientific, instructive, etc. towards the merely informative, as a result of which the copyright claim gradually peters out as it moves toward this more anodyne material. To supply the community with information, in the massive way needed nowadays, a copyright impediment would be obnoxious: the universal right to information, and also the nature itself of informative text-material hardly give any support to such a claim.

The "author" concept should not cover this too watery kind. True authorship however: articles, monographs, lectures, papers, reports (though some of these types of brainwork are group or team-created) deservedly should be remunerated on merit. However, the need for breaking up, rearranging, selectively screening, dispersing and building up into new "wholes", often does away - at least partly - with the unity and permanency of the original production. Moreover this original production speedily gets superseded by new developments or changing situations.

No. 6 - *Publishers believe that their sales rate will diminish because the use of new technologies permits libraries to reproduce quickly the deposited literature and to multiply free copy delivery in an almost uncontrollable manner. Libraries think that sales rates have not diminished through their copy delivery and that, on the contrary, they act as a kind of publicity agent for publishers.*

The link between on-line cataloguing and automated document delivery/inter library loan systems may even give scope for considerable purchases by the library world and help to increase the publishers' income.

ARGUMENTS:

6.1 - Either publishers get better copyright protection, including the right to disclosures (via independent or official inspectors) of infringement by large users, or the publishers will sink to a subsidized set of organizations - like in the communist countries - where scientific publishing "flourishes" but with huge deficits.

In royalty payments a differentiation must be made between the very low circulation specialized scientific documents and the medium run educational circulation and other material.

Data bases should also pay royalties and a publisher should have the right to know that his material is included.

6.2 - Problem of subscription to journals vs. bibliographic data bases: if a documentalist frequently discovers interesting references in a journal through the announcements made by data bases, he will certainly subscribe to the journal itself. Thus the journal has had publicity through the *secondary services*.

6.3 - The development of new possibilities of documentary searching by computer could lead to another modification of the commercial equilibrium which would no longer remain in conflict but be advantageous to both parties - to the publishers and the secondary services. Because they can be handled easily, the products of secondary services (abstract journals, bibliographies, data bases) will be used more frequently than before. In their paper format, many bibliographies used to be stored at the far end of a library without ever being flipped through. This is why *secondary services* ought to become more efficient in becoming better publicity agents to the publishers, particularly through quicker announcements.

No. 7 - *Publishers and bookshops taking advantage of new technologies could join existing, or create together their own, well functioning information networks. This could diminish the privilege of depository libraries, because it would be possible to deposit the original literature directly at the publisher's or in the bookshop, and to announce it and provide payable copy delivery through these networks.*

ARGUMENTS:

7.1 - An advantage of such a system would consist in the elimination of intermediaries and a quick distribution directly from the publisher. Through the elimination of intermediaries, however, particularly if the importance of depository libraries becomes diminished, literature as a whole might become polluted and indexing manipulated (lack of a central coordination).

7.2 - Such a system allowing direct access to the literature located at the *publisher's* or the *bookshop* would reduce the user's searching time. But it is not in the user's interest to "by-pass" the *library*, because he will miss other services which the library could render (information on the quantity of titles in a certain subject field, presentation of works in a certain subject field to the user, if abstracts would not be sufficiently informative, etc.). Even browsing in a library can be very valuable to the user. These precious services could disappear if the librarian was no longer directly in contact with his clients.

No. 8 - *The boundary between primary and secondary information may be blurred by the use of new technologies. Full text storage, for example, could make secondary services superfluous because the secondary announcement can directly be assumed by the publisher through his own networks (built-in programming for secondary data recorded and transmitted at the moment of full text input).*

ARGUMENTS:

8.1 - Secondary services have nothing to fear from new technologies, because:

- of the multiciplicity of data bases;
- of the heterogeneous character of access modes;

- better knowledge of contents will make the user even more dependent on the intermediaries;

- the more and more sophisticated forms of searching cannot be followed by the user through continuous apprenticeships, therefore he will need the intermediary.

8.2 - The new technologies allow direct access to primary sources through on-line searching, teleordering of original documents and telecopying. Thus many documentalists fear that their jobs might disappear because the users will be tempted to do the work themselves. Even large documentation centres and information brokers share these fears. But document searching necessitates the exact knowledge of the data bases concerned and of the software of the different systems.

Direct searching for primary documents will also be rendered difficult because of the language problems, and automatic translation interfaces will hardly be effectively realized in the near future.

No. 9 - *The work of secondary services could be facilitated through cooperation agreements with publishers who could transmit the secondary data to them at the earliest possible moment. The secondary information will interest different user circles and require a different marketing policy - therefore a separation of tasks between publishers and secondary services is recommended.*

ARGUMENTS:

9.1 - This assumes that the secondary services do have a marketing policy that is significantly more effective than that of the originator/publisher - there is no evidence.

9.2 - Such a cooperation does already exist for the IC Marc/UK Marc data base ("cataloguing-in-publication") whereby the publisher communicates information on the works to be published to the secondary services (Library of Congress, Blaise) directly at the moment when the print proof is issued. If teleordering was combined with such a system, a service would be rendered to publishers.

No. 10 - *A clearing-house (one per country or groups of countries) should be created to which publishers would electronically forward titles, abstracts and bibliography of their new editions. This clearing-house would disseminate the information to existing secondary services throughout the world according to the different interest profiles. The clearing-house could also introduce an automatic translation interface and communicate to the different secondary services, together with the original language data, an unrevised version of the translation into the language of their country.*

ARGUMENTS:

10.1 - This presupposes availability of suitable equipment, including software, to a majority of publishers - large and small - in addition to solving a number of other problems. Unlikely for a number of years yet.

10.2 - In France, an important source of scientific (and technical) literature exists in theses and graduate dissertations prepared in all sorts of scientific and technical educational institutions. An important part of these theses and dissertations is not published. Therefore studies on the same subject with almost identical solutions are often duplicated or even multiplied. PRIA has started some coordination by issuing an announcement bulletin and microfiches, which are however restricted to certain subjects and still very incomplete. Could this problem be resolved on a European level?

10.3 - The language and professional barrier is an important factor contributing to the difficulties of access to primary documents. New technologies do not resolve this problem. Particularly in theses, the useful information is often drowned within an enormous text of limited interest. Another example is literature written in German and concerning technology or informatics; it is mostly written in as many different jargons as there are authors, so that even professional translators have difficulties because the terminology used does not exist in any dictionary or lexicon. The third case where difficulties arise concerns patents, where the original information is completely hidden in a legal and technical jargon. In such cases, automatic translation is out of the question, and an effective documentary system is absolutely necessary.

10.4 - See Position No. 4.1 (Clearing-house collecting *legal* literature).

10.5 - See Position No. 4.2 (Clearing-house collecting *social sciences* literature).

ISSUE No. 11:

LEGAL IMPLICATIONS OF NEW PUBLISHING TECHNOLOGIES

POSITIONS

COPYRIGHT

No. 1 - *The fact that with the possibilities offered by the new publishing and distribution technologies (ranging from word processors to satellite transmission), the same information will pass through a number of technologically different channels, may lead to significant problems because unauthorized copying can hardly be controlled.*

A new copyright law should therefore be elaborated by governments in cooperation with authors, publishers, libraries and secondary services associations to guarantee to each one the appropriate return of the contribution which he gives to information and knowledge.

ARGUMENTS:

1.1 - When a symposium was organized by the Commission of the European Communities on the implementation of *Euronet*, *nothing* had been pronounced on the circulation of information in Europe, particularly the copyright question had been ignored.

a) What is the possibility of a Member State opposing the entry on its territory of information which has not been protected in the country of origin, while information of this kind is protected in the Member State concerned? Can this Member State appeal to Article 36, second sentence, and Article 85 or 86 of the Treaty of Rome?

b) If such information, in the country of origin, has been remunerated by means of a legal licence, can the Member State oppose entry on its territory by appealing to Article 234 of the Treaty of Rome which permits the appeal to article 16, paragraph 2 of the Convention of Bern (1971)?

Resuming the intellectual approach of the European Court of Justice, would it not be essential, before contemplating a reform of any kind of copyright legislation, to define particularly the "specific object" of copyright in the socio-cultural context of the given moment? This would allow the fixing of its parameters and derogations.

1.2 - Problem of copyright owner whose information is used or *accessed by a system like Euronet:*

It is clear that reproduction by photocopying or facsimile-transmission must undermine sales of publications.

- Copyright law should apply to the making of copies by any means (incl. reproduction by computer from a database); therefore a system like Euronet cannot operate without the express consent of the copyright owners when their material is being used.

- Payment should be made to the copyright owners in respect of each transaction. Those who operate the system must ensure that such payments reach the correct individual or copyright owners, though collective arrangements could be negotiated with bodies representing copyright holders to facilitate this.

- Those who operate the system must indemnify those associations of copyright owners with whom they reach agreement against action by copyright owners outside the associations.

- Reciprocal agreements must be reached with copyright owners outside the EEC.

- Any practices operated within Euronet should also be adopted for internal transactions within the Member States.

1.3 - *Suggestions for remuneration* to be given to copyright owners for the use of their material in a system such as Euronet:

- Licences may be issued on an individual or collective basis.

- Copyright owners should be able to negotiate a "lump sum" payment in respect of most works.

- Certain works which are confidential or above an agreed level of price may have to be excluded from the system or compensated by a higher fee.

- Electronically published material will also have to be charged at a higher rate.

- Fees should be charged to the ultimate user and distributed by national collecting societies.

1.4 - Development of *viewdata systems* will aggravate copyright problems beyond pure photocopying. Viewdata, using a simple audio signal, can be easily recorded on audio-tape. In Germany, France etc, viewdata is to be a "front end" distributor of data derived from on-line databases at the "publisher's" own premises. But viewdata terminals will be in mass use - 1,000,000s or more - far beyond institutional control if many of them chose to take audio-copies of material deriving from the 'base' databases.

The only solution may lie in a high initial price for access, rather than in the law.

1.5 - I am chairman of a committee or group in the United Kingdom which is concerned with the question of copyright. This group consists of representatives of four associations of publishers and four associations of authors. It is not a governmental committee. It owes its origin to a government report (the *Whitford Report on Copyright and Designs Law,* which appeared two-and-a-half years ago). That report recommended, in relation to copyright and photocopying, that a system of '*blanket licensing*' should be instituted, to safeguard the owners of copyright in this new technological situation. We are concerned to try to find, in accordance with the Whitford recommendation, a form of licensing which will be fair to all the parties concerned.

One practical step we have taken. We have mounted, with the co-operation of the educational authorities in an important part of the United Kingdom, a *sample survey of the use of photo-copying in schools.* We chose schools as a finite constituency, with local education authorities responsible for them, so that we could establish a basis of fact in an area where factual information is at present conspicuously lacking. This sample survey, covering urban and rural schools, on a statistically valid basis, is at this moment going on.

The purpose is to arrive at an agreed *pilot scheme of licensing.* Clearly, this covers only a small part of the whole field; but if we can work out something practical and acceptable - fair to everybody concerned - it may serve as an example which might be followed and extended.

1.6 - When *copyright on scientific articles in periodicals* is evoked, I think that the importance of this problem is somehow exaggerated; it concerns in fact only some *so-called* "scientific" editors whose finality is to be purely profit-making. Most learned societies publishing scientific literature do have as their unique and essential aim in their statutes the dissemination of the scientific thought of their members *within the economic limits* allowing the maintenance and continuation of their communication medium. Under these conditions, they may be satisfied if interest is shown to one of their members even in the form of a non-remunerated photocopy of his article.

1.7 - Example of the *British Library Lending Division:* the majority of the users are academic libraries - not industrial libraries. We have a great deal of evidence that academic libraries must now ration their requests to us - that problems are compounded because many of them are also being forced to cut down on acquisitions. I am sympathetic to the arguments of publishers, nevertheless I would urge caution. *Substantial royalties* may well *increase the problems of academic libraries* and increase rather than decrease the barriers to information flow. Ironically, the introduction of royalties may well cause libraries to cancel the purchase of little used material and to rely more on inter-library loan rather than photocopying.

1.8 - The importance of copyright in encouraging dissemination (actually recognized by software producers and secondary publishers) is even more important for primary publishers who have difficulty in exercising control unless there is general, legal and moral acceptance of remuneration for use of copyright material.

1.9 - The technical feasibility of controlling unauthorized copying is not the main problem. The critical issue is to enforce copyright protection in libraries directly or indirectly funded by public money and large industrial libraries, in the first instance.

No. 2 - *A legal framework should be elaborated which guarantees smooth job restructurations which become more and more necessary through continuously new developing technologies. This framework should include the continuous examination of employment elimination and new employment creation through these technologies, the possibilities for professional re-education, and possible government assistance to the publisher who may come into the difficult situation of having no other choice than to discontinue his work or to buy costly equipment which could oblige him to dismiss a part of his personnel.*

ARGUMENT:

2.1 - You cannot at the same time deprive publishers of a significant part of their income and implement "smooth job restructuration" if this means that publishers should continue to pay wages for people they have no work for.

HEALTH

No. 3 - *Thorough studies should be undertaken with regard to the health condition of personnel working with new technological publishing facilities. On the basis of these studies, legal standards should be elaborated with regard to the number of hours allowed for using such equipment and to the conditions under which such equipment is located (aeration, size of rooms, etc.).*

ARGUMENTS:

3.1 - Each technical innovation should be investigated by multidisciplinary teams composed of the equipment designers, psychiatrists, ergonomists, physicians and trade unionists, in order to realize the adaptation of equipment to man.

3.2 - As an editor, I may comparatively soon have to take decisions about investments in electronic equipment. If an investment of this type will lead to health problems for the staff, it is of little use, both humanely and economically - a calculated cost saving may lead to increased cost on another account. One needs good data on this.

TECHNICAL STANDARDS

No. 4 - *Studies should be launched to examine the degree of incompatibility between different hard and software systems and converters. These studies should serve as a basis for the elaboration of legally prescribed standards which will permit overall access to information networks.*

ARGUMENT:

4.1 - When elaborating hardware/software standards for systems and converters, one should observe that objectives which are economically realisable on a competitive basis are not hampered by bureaucratic barriers. Technical standards for equipment should be restricted to health and safety regulations and to connector equipment. All other questions can be settled by market mechanisms and ought not to be complicated by the legislator through long lasting proceedings in order to achieve some decision.

SMALL BUT EFFICIENT PUBLISHERS

No. 5 - *A thorough study should be launched to examine the situation of small publishers who, due to the nature of their publishing programme and to their financial situation, are not able to move into new publishing technologies or prefer to continue a traditional production in order to maintain the quality of their publications.*

The outcome of this study should constitute the basis for the elaboration of a legal framework which would safeguard the interests of the individual traditional publisher and simultaneously maintain the diversity and quality of information in general.

ARGUMENTS:

5.1 - The leading secondary services have maintained their quality control jumping into tape and on-line transfer. Why should it not be possible to maintain this control using new systems for other types of information transfer.

Using new techniques for information transfer, I suppose much attention has to be given to the *structure of the information* supplied through new media. Publishers and information specialists have a task to convince authors (mainly technical and scientific ones are meant here) to *limit their information message* to the real information content.

Publishers have to become aware of the fact that they have a duty in this respect for the sake of the information they handle. It could well be that they have to do so for the sake of their own survival

5.2 - It is important to *maintain the variety of input* by ensuring the role of competing publishers in new complex, electronic based, high capital systems. This is of vital democratic and scientific interest. A work reporting a factual experiment now may be valuable in 5 years' time because of other developments.

5.3 - *Protection of small publishers is essential.* The majority of scientific publications are realised by the "amateurs" of the publishing world (universities, learned societies): they assume the largest part of information transfer. The discernment by the referee as an intermediary is more efficient in small journals than in journals printed by large publishing houses where articles though of lower quality are accepted even after having been rejected by the small journals.

5.4 - During this symposium, a speaker stated that in the future new technologies would certainly give rise to a *restructuration of the journal market.* This constitutes a particular threat to specialized journals in the scientific and technical field. Publishers finance their journals mostly through advertising. This is why some commercial publishers tend more and more to publish articles whose subjects are close to business practice. The specialized technical/scientific journals have got to (or want to, according to their statutes) publish scientific works which do not generate commercial advertisements. If the publishers of these journals would have to turn away more and more from such behaviour which involves "heavy losses", a damage to the branch of science concerned will be inevitable. This is why the burden which new technologies may lay on these small scientific publishers must be well considered.

5.5 - Because of their direct contact (panels, working groups, conferences, letters, etc.) with authors, information producers and information users, the "learned societies", for example, as publishers and editors of specialized scientific/technical journals are one of the most important intermediaries for the communication of reliable information and data. These societies have the possibility to select good quality information which would require a corresponding quality of the consecutively used storage media (journals, indexes). Wouldn't the new technologies used for data collection, storage and dissemination which will have to be operated on a commercial basis lead to a reduction of quality because the specialist knowlege of the "learned societies" would possibly be by-passed?

In my opinion, such a development is within sight, because the small scientific/technical socieities will hardly be in a financial position to continue publishing their specialised media. In Germany, for example, this would concern the so-called "gemeinnützige Vereine" (non-profit making associations of public interest) who depend on the financial benevolence of their members and are not allowed to work on a commercial basis.

The EC Commission should provide in time for such measures which guarantee the continuation of the incontestably important cooperation of the "learned societies" on the information market and within the structure of modern information technologies.

5.6 - Many users of medical databanks prefer to have a small selection from recognized core journals with high quality referee systems rather than all available data. For me as an author it means that "my" journal should continue to check data as well as it can in cooperation between authors, referees and editors.

QUALITY, CORRECTNESS and DIVERSITY of INFORMATION

No. 6 - *The new publishing technologies allow information, which may be originating from one single or a few persons, to be disseminated throughout the world in a few minutes. Information can therefore easily be monopolized, influenced and manipulated. False information may be disseminated in seconds and corrected only years after. Long running studies may be based on initially wrong information. Precious information stored by modern media can be destroyed forever through sabotage or terrorism. It is necessary to undertake thorough studies into the possibilities of information misuse or destruction through modern technologies and to elaborate a legal framework which safeguards the continuity of free, correct and diversified information for everybody.*

ARGUMENTS:

6.1 - While the symposium has been very much concerned with the dissemination of information the reverse side of the coin, that of *confidentiality,* has to be considered. With the enormous growth of computerized on-line services more and more restricted data can be "raided". There should be an urgent review of the law in this area in order to give the best possible protection. There should be severe penalties for infringements, criminal as well as civil.

6.2 - It is unclear how the quality control of the input can be exerted - i.e. the control of basic correctness of facts - given that it is at present exerted by the refereeing and editing systems of the internationally accepted journals. This is an important question since far-reaching decisions may well be taken on the basis of the content of the data banks, and the decisions can then never be better than the quality of the data.

6.3 - *An "inventory of the types of communication existing in the informatics age"* should be established. It should comprise:

a) A catalogue of the different types of communication;

b) An economic and sociological study of their cross-relations;

c) *What* to communicate to *whom* (type and number of *users*), in which form and on which support;

d) Category of data: factual - notional - permanent - transitory;

e) Category of users: literate or illiterate, active or passive;

f) Category of media: sound - image - code (graphics) - symbol (ideogram);

g) Category of support: electronic - interactive - static - tape - film - paper;

h) Possibility of creating an automatic model for ponderation of the variables of these different elements.

Such an inventory could serve as a starting point to determine the possibilities and dangers inherent in the different existing and future communication systems. It could help to preserve the manifoldness of information and it would permit elaboration of the suggested *legal framework* on a realistic basis.

6.4 - Concerning the *diversity of information,* it should also be considered that 80% of the world population are illiterate. An example is the Arabic world where the proportion of people who are able to read their language is relatively small. (To a great extent, this is due to the syllabic structure of the language and the almost total absence of vowel signs.) This problem is dramatic for all technical literature where vowels do have to be used. New technologies can bring a solution to this very old problem which goes back to the invention of printing and has not yet been solved. A meeting on this problem will take place in November 1980 in Fontenant by the RECI Centre in cooperation with *LURE*, UNESCO and the French Imprimerie Nationale and the results will have to be followed.

6.5 - The suggested *legal framework* should particularly comprise legal harmonization in the following sectors:

a) Copyright and literary copyright;

b) Law applying to right for re-statements and counter-statements ("Gegendarstellungsrecht") and to the right for protection of the personality in case of prejudice through new media;

c) Printing law and analogue regulations in the case of new media;

d) Penal legislation applying to the press ("Pressestrafrecht") extended to new media including specific marginal regulations (e.g. right to refuse testimony) relating to law suit.

This "legal framework" should possibly not only apply to the EC jurisdiction but be extended to the territory of the Council of Europe Member States.

List of Participants

ALY, H,	GDMB Gesellschaft Deutscher Metallhütten- und Bergleute Postfach 210 D - 3392 CLAUSTHAL-ZELLERFELD
ANDERLA, G	Commission of the European Communities Directorate-General XIII-B Scientific and Technical Information and Information Management Bâtiment Jean Monnet Kirchberg LUXEMBOURG
APPLEYARD, R.	Commission of the European Communities Directorate-General XIII Scientific and Technical Information and Information Management Bâtiment Jean Monnet Kirchberg LUXEMBOURG
ARCHIMBAUD, C. (Mme)	E.D.F. 1 Avenue du Général de Gaulle F - 92141 CLAMART
ASSELBORN, J.-C.	Centre Universitaire Boulevard Salentiny 59 LUXEMBOURG
ATHERTON, M.J.	British Post Office 88 Hills Road CAMBRIDGE CB2 1PE England
AUBURTIN, M.	L'Argus Journal International des Assurances 2 rue de Châteaudun F - 75441 PARIS Cedex 09
BAIVY, J.M.	Technical Translation International 119 Chee. St. Pierre B - 1040 BRUXELLES

BARBOT, R.	O.C.D.E. 2 rue André-Pascal F - 75775 PARIS Cedex 16
BARLOW, D.H.	INSPEC Savoy Place LONDON WC2R 0BL England
BARNIG, M.	Administration des Postes et Telecommunications Boîte Postale 2061 LUXEMBOURG
BARTHÉLEMY, J.-P.	11 Avenue Victor Hugo F - 75116 PARIS
BATTRICK, B.	European Space Agency Zwarteweg NOORDWIJK 2200 AG The Netherlands
BEENEN, H.	European Patent Office P.B. 5818 Patentlaan 2 2280 HV RIJSWIJK (ZH) The Netherlands
BERARD, J.	Telesystems Rue du Cherche-Midi F - 75015 PARIS
BERGER, A.	Service juridique de la Commission C.E. Avenue de la Joyeuse Entrée, 23/27 B - 1040 BRUXELLES
BERGMANS, L.	Elsevier Sequoia S.A. P.O. Box 851 CH - 1001 LAUSANNE 1
BETHERY, M.	Bibliothèque Publique d'Information Centre Georges Pompidou F - 75191 PARIS Cedex 4

BILBOUL, R.R.	Learned Information (Europe) Ltd. Besselsleigh Road ABINGDON Oxford OX13 6EF England
BJØRN, B.	260 Rue Principale Munsbach G.D. de LUXEMBOURG
BLESBOIS, J.-P.	Syndicat National de l'Edition 117 Boulevard Saint-Germain F - 75006 PARIS
BLOK, H.	GTE ATEA Industriepark Klein Gent B - 2410 HERENTALS
BÖCKING, B.	Eurograf Square Ambiorix 40 B - 1040 BRUXELLES
BONNET, P.	Commission des Communautes Européennes SEPC 200, rue de la Loi B - BRUXELLES
BORNES, C.	I.R.I.A. Domaine de Voluceau B.P. 105 F - 78150 LE CHESNAY
BOTHE, A.	Ministry of Foreign Affairs Casmariestr. 16 The HAGUE The Netherlands
BOUVIER, M.	Direction Generale Telecommunications La Villereine - 3 Allee Kurzenne F - 78530 BUC
BRADLEY, C.	The Publishers Association 19 Bedford Square LONDON WCB 3HJ England

BREE, R.	An der Herrenwiese 6 D - 5330 KÖNIGSWINTER 41
BREEKWEG, C.	N.V. ICV - Belgie Louizalaan 485 B - 1050 BRUXELLES
BREIDENSTEIN, H.J.	Umschau Verlag Stuttgarter Str. 18-24 D - FRANKFURT 101
BULLENS, O.	European Patent Office P.B. 5818 Patentlaan 2 2280 HV RIJSWIJK (ZH) The Netherlands
BURNHAM, H.	Parlément Européen Centre Européen - Kirchberg LUXEMBOURG
BUSACHI, F.	Commission des Communautés Européennes Berlaymont 2/116 B - 1049 BRUXELLES
CAMERON, J.	John Wiley & Sons Limited Baffins Lane CHICHESTER Sussex PO19 1UD England
CAMPION-VINCENT, V. (Mme)	Centre National de la Recherche Scientifique 54 boulevard Raspail F - 75006 - PARIS
CASASSA, G.	SEAT - Società Elenchi Ufficiali abbonati al Telefono p.a. Via Aurelio Saffi 18 I - 10138 TORINO
CHERKI, C.	57 rue de Seine F - 75006 PARIS
CHRISPEELS, O.	Ced - Samson S.A. 7 Rue Ph. de Champagne B - 1000 BRUXELLES

CLEDES, P.-P.	Centre de Recherches, d'Information et de Documentation Notariales (CRIDON) 59 bis rue de Créqui F - 69452 LYON Cedex 3
COGHLAN, M.	Institute for Industrial Research & Standards Ballymun Road DUBLIN Ireland
COLBACH, A.	Luxemburger Wort 2 rue Christophe Plantin LUXEMBOURG
COOPER, M.	Quantum Science Corporation 16 Charles II Street LONDON SW1Y 4QU England
CORNELIUS, P.	Georgenschwaigstr. 11 D - 8000 MÜNCHEN 40
CORNISH, D.	The British Council Hahnenstr. 6 D - 5000 KÖLN 1
CREMER, M.	Bischofsweg 31 D - 6000 FRANKFURT/MAIN
CURRY, J.P.	Intermediair N.Z. Voorburgwal 225 1012 RL AMSTERDAM The Netherlands
DANET, A.	Honeywell Bull 68 route de Versailles F - 78 LOUVECIENNES
D'ATTILIA, G.L.	SEAT Ufficio Studi di D.G. Via Aurelio Saffi 18 I - 10138 TORINO
DAVIDS, H.	Elsevier - NDU Jan van Galenstraat 335 AMSTERDAM The Netherlands

DAVIES, G.	Commission of the European Communities Directorate-General XIII-B Scientific and Technical Information and Information Management Bâtiment Jean Monnet Kirchberg LUXEMBOURG
DAWSON, P.M.	Ministry of Agriculture Agricultural Development & Advisory Service Great Westminster House Horseferry Road LONDON SW1P 2AE England
DE GREVE, M.	A.I.M.A.V. Blandijnberg 2 B - 9000 GENT
DE HALLEUX, N.	CONSEIL DE L'EUROPE Palais de l'Europe F - 67006 STRASBOURG
DE LA FABRÈGUE, F. (Mme)	I.R.I.A. Domaine de Voluceau B.P. 105 Rocquencourt F - 78150 LE CHESNAY
DE LUZE, G.	LAROUSSE 17 rue du Montparnasse F - 75280 PARIS Cedex 06
DEMAEGHT, J.	Cour de Justice des Communautés européennes Plateau du Kirchberg LUXEMBOURG
DERMINEUR, B.	Cercle de la Librairie Bd. St. Germain 117 F - 75006 PARIS
DEUBEL, A.	Librairie Lavoisier Editions Technique et Documentation 11 rue Lavoisier F - 75384 PARIS Cedex 08

DE VRIES, H.	European Patent Office Rosenheimerstr. 30 D - 8000 MÜNCHEN 80
DIDIER, P.	Association Belge des Editeurs de Langue Française Avenue du Parc 111 B - 1080 BRUXELLES
DREYER, R.	Office des Publications C.E. Avenue de la Joyeuse Entrée 23/27 B - 1040 BRUXELLES
DRUBBA, H.	Universitätsbibliothek und Technische Informationsbibliothek Hannover Welfengarten 1 B D - 3000 HANNOVER 1
DUTTA, S.K.	Indian Space Research Organisation, Bangalore ISRO HQ. Cauvery Bhavan District Office Road BANGALORE 560052 India
EMRINGER, L.	Office des Publications des Communautés Européennes 5 Rue du Commerce LUXEMBOURG
EVANS, A.G.J..	The Biochemical Society 7 Warwick Court High Holborn LONDON WC1R 5DP England
FAUCHÈRE, P.-M.	Institute des Sciences de l'Information et de la Documentation Université de Clermont II Rue de la Pras Saint Bonnet-de-Rochefort F - 03800 GANNAT
FAVRE BULLE, B.	G.E.D.I.M. 19 rue du Grand-Moulin F - 42029 SAINT-ETIENNE Cédex

FAYET, J.C.	Institut National des Industries et Arts Graphiques 296 rue de Charenton F - 75012 PARIS
FAZZINO, E.	Commission of the European Communities 16 rue Hydraulique B - BRUXELLES
FELDES, C.	S.G.2 124B rue du Kiem LUXEMBOURG
FERRIS, P.	John Wiley & Sons Limited Baffins Lane CHICHESTER Sussex PO19 1UD England
FERRY, G. (Mrs)	New Scientist King's Reach Tower Stamford Street LONDON S.E.1. England
FISCHER, J.	Hahn-Meitner-Institut für Kernforschung Berlin GmbH Glienicker Strasse 100 D - 1000 BERLIN 39
FLEURENT-DIDIER, J.	Groupe Audiovisuel de l'Edition Syndicat National de l'Edition 117 Boulevard Saint-Germain F - 75006 PARIS
FOGELBERG, P.	University of Helsinki Hallituskatu 11-13 SF-00100 HELSINKI 10 Finland
FOLSCHETTE, E.	Imprimerie Centrale 15 rue du Commerce LUXEMBOURG
FORREST, A.	Council of the European Communities 170, Rue de la Loi B - 1048 BRUXELLES

FOUWELS, A.	Commission of the European Communities 200, Rue de la Loi B - BRUXELLES
FOX, J.A.	NFER Publishing Company Limited The Mere Upton Park SLOUGH Berks. SL1 2DQ England
FRANKLIN, J.	Elsevier/North-Holland P.O. Box 1527 1000 BM AMSTERDAM The Netherlands
FREDRIKSSON, E.M.	North-Holland Publishing Company Jan Van Galenstr. 335 AMSTERDAM W. The Netherlands
GARDINER, K.	Office des Publications des Communautés Européennes 5 Rue du Commerce LUXEMBOURG
GASKELL, E.	Commission of the European Communities D.G. IX 200, Rue de la Loi B - BRUXELLES
GATES, Yuri, (Mrs)	PIRA Randalls Road LEATHERHEAD Surrey KT22 7RU England
GEBHARD, R.	CIBA-GEIGY AG WD/Bibliothekswesen Schwarzwaldallee 215, Rosental CH - 4002 BASEL
GEISEN, B.	Kalle, Niederlassung der Hoedist AG Postfach 3540 D - 6200 WIESBADEN 1

GERLACH, L. (Miss)	Commission of the European Communities Directorate-General XIII-A Scientific and Technical Information and Information Management Bâtiment Jean Monnet Kirchberg LUXEMBOURG
GIACCARI, M.D.	Commission of the European Communities 88 Avenue Louise B - 1050 BRUXELLES
GIBB, J.M.	Commission of the European Communities Directorate-General XIII-A Scientific and Technical Information and Information Management Bâtiment Jean Monnet Kirchberg LUXEMBOURG
GILMAN, P.	S.A. Imprimerie et Journal LA MEUSE Boulevard de la Sauvenière 8-12 B - 4000 LIÈGE
GINESTIE, J. (Mme)	40 rue de Monceau F - 75008 PARIS
GINESTIE, P.	40 rue de Monceau F - 75008 PARIS
GODAERT, P.	Professeur à l'Université de Louvain 103 Route du Vin SCHENGEN G. Duché de Luxembourg
GOEBEL, J.W.	Gesellschaft für Information und Dokumentation Lyoner Str. 44-48 D - 6000 FRANKFURT/MAIN 71
GOFFART, B.	Institut National des Industries Extractives 200 rue du Chera B - 4000 LIÈGE
GORDON, M.	7-9 rue Emile Dubois F - PARIS XIV

GORMEZANO, R.	S.G.2 Vion Whitcomb F - 25016 PARIS
GÖTTE, L.J.	N.V. Kluwer Postbox 23 DEVENTER The Netherlands
GRAHAM, G.	Butterworth & Co. (Publishers) Ltd. 88 Kingsway LONDON WC2B 6AB England
GREINER, S.	Bundessprachenamt Horbeller Str. D - HUERTH
GROS, E.	Scientific Information Consultant 661 Finchley Road LONDON NW2 2HN England
GRÜNHAGEN, H.	Training Department for Booksellers and Publishers Broerelaan 6 EINDHOVEN The Netherlands
GRUTZMACHER, E.	Scientific and Technical Documentation and Information Centre for the Armed Forces Van den Burchlaan 31 2597 PC THE HAGUE The Netherlands
GUCKLER, G.	Max Hueber Verlag Krausstr. 30 D - 8045 ISMANING/MÜNCHEN
HANCOCK, L.	National Children's Bureau 8 Wakley Street Islington LONDON EC1V 7QE England
HANISCH, G.	Deutsche Bundesbank Postfach 2633 D - 6000 FRANKFURT 1

HARMSEN, D.-M.	Institut für Systemtechnik und Innovationsforschung Sebastian-Kneipp-Strasse 12-14 D - 7500 KARLSRUHE 1
HAUGG, N.	Deutsches Patentamt Zweibrückenstrasse 12 D - 8000 MÜNCHEN 22
HAUSER, W.	Physikalisch-Technische Bundesanstalt Bundesallee 100 D - 3300 BRAUNSCHWEIG
HEBRANT MACEVICIUS, M. (Mme)	Parlément Européen Bibliothèque Centre Européen - Kirchberg LUXEMBOURG
HEDDESHEIMER, F.	Bibliothèque Universitaire de Nancy 49 rue de la Libération F - NANCY
HEINTZE, H.J.	ECOFOOD GmbH Reindorf 12 D - 2110 BUCHHOLZ i.d. Nordheide
HEINZE, U.	WAZ - Zeitungsgruppe Essen Derner Strasse 8 D - 4618 KAMEN
HEMMING, P.	Ernst Klett Verlag Rotebühlstrasse 77 D - 7000 STUTTGART 1
HETTRICH, E.	Commission of the European Communities Directorate-General XIII-A Scientific and Technical Information and Information Management Bâtiment Jean Monnet Kirchberg LUXEMBOURG
HILLS, P.J.	Primary Communications Research Centre University of Leicester LEICESTER LEI 7RH England

HJORT, B.	Gyldendal Lindgrens Alle 12 COPENHAGEN Denmark
HOUTART, E.	C.R.E.D.O.C. 34 de la Montagne B - 1000 BRUXELLES
HUBER, R.	Kommission der Europäischen Gemeinschaften D.G. XII-FAST 200 rue de la Loi B - 1049 BRUXELLES
HUNAULT, M. (Mme)	Bureau de presse et d'information des Communautés européennes 61 rue des Belles-Feuilles F - 75116 PARIS
ICART, C. (Mme)	Federation Francaise de l'Imprimerie 115 Bd. Saint-Germain F - 75006 PARIS
IRGENS-JENSEN, S.	Nordic Publishing Board in Science Munthes GT.29 OSLO 2 Norway
JACKISCH, W.	Verlag Glückauf GmbH Postfach 1794 D - 4300 ESSEN 1
JAY, B.	Commission of the European Communities Directorate-General XIII-A Scientific and Technical Information and Information Management Bâtiment Jean Monnet Kirchberg LUXEMBOURG
JEWITT, C.	Oriel Computer Services Limited 1 & 3 West Street CHIPPING NORTON Oxon OX7 5LY England

JUCQUOIS, M. (Mme)	Forschungsabteilung für Dok. und Inform. Universität Düsseldorf Wersderer Dorfstrasse 122 D - 4000 DÜSSELDORF 1
KEMME, G.	Stichling Moderne Media Herlengracht 257 AMSTERDAM The Netherlands
KEMMERICH, M. (Frau)	Kernforschungszentrum Karlsruhe GmbH Postfach 3640 D - 7507 KARLSRUHE
KLEIN, K.	Verlag Glückauf GmbH Postfach 1794 D - 4300 ESSEN 1
KNOCHÉ, G.	Documentation Française 31 Quai Voltaire F - 75340 PARIS Cedex 07
KORVING, F.	N.V. Nederlanse Spoorwegen Moreelsepark 1 3511 EP UTRECHT The Netherlands
KRAMER, H.E.	Deutsches Patentamt Zweibrückenstrasse 12 D - 8000 MÜNCHEN 22
KUMMER, H.	Bundesdruckerei 91 Oranienstr. D - 1000 BERLIN 61
KYLIN, A.	Physiologia Plantarum University of Lund Fack S-220 07 LUND, 7 Sweden
LACHMEIJER, G.	Gerard Doustraat 28 UTRECHT The Netherlands
LAMY, G.	Sté LAMY S.A. 155 Rue Legendre F - 75017 PARIS

LANNOY, J.	Commission of the European Communities Directorate-General XIII-A Scientific and Technical Information and Information Management Bâtiment Jean Monnet Kirchberg LUXEMBOURG
LEHOUCK, P.	Ministère des Affaires Etrangères 2 Rue des Quatre Bras B - 1000 BRUXELLES
LEPELLETIER, M. (Mme)	Commission of the European Communities Avenue Pasteur 20 LUXEMBOURG
LEPRÊTRE, J.-P.	Bureau de Recherches Géologiques et Minières B.P. 6009 F - 45060 ORLEANS
LEUTENEGGER, W.	Ciba - Geigy AG K 1370 C CH - 4000 BASEL
LINK, W.	Max Hueber Verlag Krausstrasse 30 D - 8045 ISMANING b. MÜNCHEN
LOFF, I. (Mme)	Secretariado para a Integraçao Europeia Almeida Brandao, 7 1200 LISBOA Portugal
LOHNER, F.	Gesellschaft für Information und Dokumentation (GID) Lyoner Str. 44-48 D - 6000 FRANKFURT/MAIN 71
LOVEDAY, A.D.	World Health Organization Avenue Appia CH - 1211 GENÈVE 27
LUBBERHUIZEN, J.-W.	Industriegroep Grafische Bedrijven KLUWER B.V. Arnhemsestraatweg 8 6881 NG VELP (Gld.) The Netherlands

LUSTAC, S.	Commission of the European Communities Directorate-General XIII-B Scientific and Technical Information and Information Management Bâtiment Jean Monnet Kirchberg LUXEMBOURG
MAASDAM, J.	Informatie en Communicatie Unie (ICU) Postbus 1115 8011 CT ZWOLLE The Netherlands
MARTIN, P.	Centre d'Etudes Vallourec Rue de Leval F - AULNOYE AYMERIES
MASLIN, J.	PIRA Randalls Road LEATHERHEAD Surrey England
MASTRODDI, F.	Commission of the European Communities Directorate-General XIII-B Scientific and Technical Information and Information Management Bâtiment Jean Monnet Kirchberg LUXEMBOURG
MAURICE, M.	Commission of the European Communities Directorate-General XIII-A Scientific and Technical Information and Information Management Bâtiment Jean Monnet Kirchberg LUXEMBOURG
McCALL, R.	GID-SfS Werderstrasse 35 D - 6900 HEIDELBERG
MEINANDER, M.	Zoologiska Museet N. Järnvägsgätan 13 SF - 00100 HELSINGFORS 10 Finland

MELLIGARI	S.G.2 54 Avenue de la Liberté LUXEMBOURG
MESNIL, B.	Direction Générale des Télécommunications 92 Bd. du Montparnasse F - PARIS 14e
MICHAUD, P.	I.N.S.E.E. 18 Boulevard Adolphe Pinard F - 75014 PARIS Cedex 14
MICHEL, J.H.	Université Libre de Bruxelles Faculté Philosophie et Lettres Avenue F.-D. Roosevelt 50 B - 1050 BRUXELLES
MICHEL, M. (Mme)	CNET - DAII Ministère des P et T 38-40 rue du Général Leclerc F - 92131 ISSY LES MOULINEAUX
MICHELET, P.	Conseil dė l'Europe Palais de l'Europe F - 67006 STRASBOURG
MIGNOT, D.	Université de Paris XIII 10 Cité St. Exupery F - 93440 DUGNY
MINDERHOUT, P.C.	nv ICU, Informatie en Communicatie Unie Stadhoudersplein 1 2404 BE ALPHEN AAN DEN RIJN The Netherlands
MORINA, D.	Parlément Européen Centre Européen - Kirchberg LUXEMBOURG
MORIZUR, J.-P.	Mission Interministérielle de l'Information Scientifique et Technique (MIDIST) 10 rue Crillon F - 75194 PARIS Cedex 04
MULLENDERS, J.	Université de Liège 11 Bois des Chevreuils B - 4040 TILFF

MULLER, G.	Commission des Comunautes Européennes 200 rue de la Loi B - 1049 BRUXELLES
MÜLLER, J.	REFA Verband für Arbeitsstudien und Betriebsorganisation e.V. Postfach 4138 D - 6100 DARMSTADT
MULLER, M.-P.	Paroisse N.D. du Rosaire 16 rue Hansel B - 6798 AUBANGE
MYATT, A.G.	British Library Lending Divison Boston Spa WETHERBY West Yorkshire LS23 7BQ England
NEJMAN, E. (Mrs)	King's College London The Strand LONDON WC2 England
NEUMANN, A. (Frau)	Infobrief Research and Technology Tanneweg 10 D - 5307 WACHTBERG-NIEDERBACHEM
NEUMANN, P.	SDV Saarbrücker Druckerei und Verlag GmbH Halbergstrasse 3, Postfach 442 D - 6600 SAARBRÜCKEN
NEUMANN, S.	Infobrief Research and Technology Tanneweg 10 D - 5307 WACHTBERG-NIEDERBACHEM
NICOLAY, D.	Commission of the European Communities Directorate-General XIII-A Scientific and Technical Information and Information Management Bâtiment Jean Monnet Kirchberg LUXEMBOURG

NIJHOFF ASSER, P.	International Group of Scientific, Technical and Medical Publishers (STM) Keizersgracht 462 1016 GE AMSTERDAM The Netherlands
NOEL, J.	Université de Liège Allée de la Cense Rouge 50 B - 4200 OUGREE
O'DONNEL, P.	Commission of the European Communities Directorate-Generale XIII-B Bâtiment Jean Monnet Kirchberg LUXEMBOURG
OLSEN, T.	Odense Universitetsbibliotek Campusvej 55 5230 ODENSE M Denmark
ORANI, M.	Commission of the European Communitie Direction-General XIII-B Bâtiment Jean Monnet Kirchberg LUXEMBOURG
OTTEN, K.	1925 Trebein Road XENIA Ohio 45385 U.S.A.
OTTO, G.	JBM Pascallstr. 100 D - 7000 STUTTGART
PACHKEVITCH, N.	Connaissance Par l'Image 15 rue Clareyron F - 75008 PARIS
PAGE, G. (Miss)	Pageant Publishing 5 Turners Wood LONDON NW11 England

PAGE, J.	International Institute for Applied Systems Analysis (IIASA) Schlossplatz 1 A-2361 LAXENBURG Austria
PATERSON, A.	LOGICA Ltd. 64 Newman Street LONDON W1A 4SE England
PATESSON, R.	Université Libre de Bruxelles Faculté de Philosophie et Lettres Avenue F.-D. Roosevelt 50 B - 1050 BRUXELLES
PEGG, M.	University of Birmingham P.O. Box 363 BIRMINGHAM B15 2TT England
PELTRE, A.	CII - H. Bull 129 Bd. d'Aulnay F - 93250 VILLEMAMBLE
PERRIER, J.F.	Ambassade du Canada 6 rue de Loxum B - 1000 BRUXELLES
PERRY, D.	Office for Official Publications of the European Communities 5 Rue du Commerce LUXEMBOURG
PHILLIPS, E.	Commission of the European Communities Directorate-General XIII-A Scientific and Technical Information and Information Management Bâtiment Jean Monnet Kirchberg LUXEMBOURG
PICHON, F.	La Revue de Métallurgie 5 rue Paul Cézanne F - 75008 PARIS

PIJNENBURG, J.	Verbindingstraat 24 B - 2540 HOVE
PILLERAULT, C.	Editions TESTS s.a. 71 bd. Richard-Lenoir F - 75011 PARIS
PION, M.	Bethoux, Burner et Associés 32 Avenue du Vercons F - MEYLAN
PIRARD, R.	I.B.M. Square Victoria Regina B - 1030 BRUXELLES
POLLITZER, R.	King's College London The Strand LONDON W.C.2 England
POSWICK, R.-F.	Abbaye de Maredsous 11 rue de l'Abbaye B - 5642 DENÉE
POUVESLE, G. (Mme)	Laboratoires d'Electronique et de Physique Appliquée (L.E.P.) 3 Avenue Descartes, B.P. 15 F - 94450 LIMEIL-BREVANNES
PRAVDIC, N.	Referral Centre of the University of Zagreb Trg Marsala Tita 3/III 41001 ZAGREB Jugoslavija
PSZENNY. F. (Mlle)	Institut National des Industries et Arts Graphiques (I.N.I.A.G.) 296 rue de Charenton F - 75012 PARIS
RAATS, M.	Federation Belge des Industries Graphiques 20 rue Belliard, B.P. 16 B - 1040 BRUXELLES
RAWLINS, C.J.	IPC Science and Technology Press Ltd. P.O. Box 63

	Westbury House Bury Street GUILDFORD Surrey GU2 5BH England
REANEY, D.	OYEZ Communications Group Norwich House 11/13 Norwich Street LONDON EC4A 1AB England
REID, A.S.	The European Parliament Centre Européen - Kirchberg LUXEMBOURG
REMKE, R.	Sonderforschungsbereich 7 Uni Münster Salzstrasse 41 D - 4400 MÜNSTER
RICHERI, G.	Regione Emilia Romagna viale Silvani 6 BOLOGNA Italy
RIEDINGER, L.	Editions Techniques pour l'Automobile et l'Industrie (E.T.A.I.) 20 Rue de la Saussière F - 92100 BOULOGNE BILLANCOURT
ROLLING, L.	Commission of the European Communities Directorate-General XIII-B Scientific and Technical Information and Information Management Bâtiment Jean Monnet LUXEMBOURG
RONCHI, L.	Commission of the European Communities Office Statistique Bâtiment Jean Monnet Kirchberg LUXEMBURG

ROSSEEL, E.	Association Internationale pour la Recherche et la Diffusion des Méthodes Audo-visuelles et Structuro-globales (AIMAV) Blandijnberg 2 B - 9000 GENT
ROTONDO', P.P.	Commission of the European Communities Directorate-General XIII-A Scientific and Technical Information and Information Management Bâtiment Jean Monnet Kirchberg LUXEMBOURG
ROTH, G. (Frau)	GID/SfS Untere Neckarstr. 19 D - 6900 HEIDELBERG
RUECKERT, D.	Commission of the European Communities Direction Générale XIII-A-1 Bâtiment Jean Monnet Kirchberg LUXEMBOURG
RUPPERT, J.	S.G.2 54 Avenue de la Liberté LUXEMBOURG
RUTGERS, A.	Centre for Agricultural Publishing and Documentation 17 Marijkeweg P.O. Box 4 6700 AA WAGENINGEN The Netherlands
SALMON, C.	LABORELEC B.P. 11 B - 1640 RHODE-ST-GENESE
SAUR, K.G.	K.G. Saur Verlag München Pössenbacherstr. 2b D - 8000 MÜNCHEN 71
SCAILQUIN, M.-C. (Mme)	Office Belge du Commerce Extérieur Boulevard Emile Jacqmain B - 1000 BRUXELLES

SCHERFF, H.	Commission of the European Communities Directorate-General XIII-A Scientific and Technical Information and Information Management Bâtiment Jean Monnet Kirchberg LUXEMBOURG
SCHILTZ, P.	Professeur éditeur Pl. du Rotenberg 3 B - EUPEN 4700
SCHMIT, A. (Mme)	Association Européenne du Ciment 2 Rue Saint-Charles F - 75740 PARIS Cedex 15
SCHNEIDER, K.W.	N.V. PHILIPS Vestdijk 2D 5600 MD EINDHOVEN The Netherlands
SCHNELL, F.	Commission of the European Communities 200 Rue de la Loi B - BRUXELLES
SCHRÖDER, K.	Kernforschungsanlage Jülich GmbH Postfach 1913 D - 517 JÜLICH
SEIDEL, M.	Börsenverein des Deutschen Buchhandels e.V. Grosser Hirschgraben 17/21 Postfach 2404 D-6000 FRANKFURT AM MAIN 1
SELLIER, A.L.	J. Schweitzer Verlag Geibelstrasse 8 D-8000 MÜNCHEN 80
SEMTURS, F.	IBM Watsonweg 2 UITHOORN The Netherlands
SIDET, A.	Commission of the European Communities D.G. X 200 Rue de la Loi B - BRUXELLES

SÖFFNER, W.
Deutsche Bundesbank
Postfach 2633
D - 6000 FRANKFURT 1

SOMMER, H.
Mannesmann Aktiengesellschaft
Postfach 5501
D - 4000 DÜSSELDORF 1

ST. AUBYN, J.D.
Institution of Electrical Engineers
P.O. Box 8
Southgate House
STEVENAGE
Herts. SG1 1HQ
England

STEPHENS, A.D.W.
Prestel International
Telephone House
Temple Avenue
LONDON EC4Y 0HL
England

STERN, B.T.
Excerpta Medica
P.O. Box 1126
1000 BC AMSTERDAM
The Netherlands

STILLER, S.
Rechenzentrum
Uni Münster
Roxeler Strasse 60
D - 4400 MÜNSTER

STREIL, J.
Cour de Justice des Communautes Européennes
Kirchberg
LUXEMBOURG

SUCHARD, J.
14 Sq. Adanson
F - 75239 PARIS Cedex 05

SUMMERS, D.
Butterworth & Co. (Publishers) Ltd.
Borough Green
SEVENOAKS
Kent TN15 8PH
England

TERMENS, C.
Direction Generale des Telecommunications
33 Avenue du Maine
F - PARIS

TERWAN, J.	Unilever Research Vlaadingen P.O. Box 114 3130 AC VLAARDINGEN The Netherlands
THALBERG, K.R.	University of Trondheim Högskoleringen 1 7034 TRONDHEIM The Netherlands
THÉDORE, J.-M.	BIU Nancy Droit 11 Place Carnot F - 54000 NANCY
THILL, P.	Impr. St. Paul (Luxemburger Wort) 5 rue M. Kirsch LUXEMBOURG
THIRION, G.	Bibliothèque Universitaire de Nancy 11 Place Carnot F - NANCY
TONELOTTO, M.	Parlément Européen Centre Européen - Kirchberg LUXEMBOURG
TONHOFER, F.	Office des Publications des Communautés Européennes 5 Rue du Commerce LUXEMBOURG
TRAKSEL, J.A.	Inter Documentation Co. Hogewoerd LEIDEN The Netherlands
TULETT, P.A.	The Institute of Physics Techno House Redcliffe Way BRISTOL BS1 6NX England
URQUHART, J.A.	University Library Newcastle upon Tyne NEWCASTLE UPON TYNE NE1 7RU England

VAN BERKEL, P.L.M.	European Systems Partners (ESPA) 107 rue de la Loi B - 1040 BRUXELLES
VAN EYSINGA, F.	Martinus Nijhoff Lange Voorhout 2501 CN THE HAGUE The Netherlands
VAN HALM, J.	Johan Van Halm & Associates P.O. Box 688 3800 AR AMERSFOORT The Netherlands
VAN LEEUWEN, J.K.W.	Excerpta Medica Keizersgracht 305 AMSTERDAM The Netherlands
VAN PROOIJEN, A.	B.E.T.A. Maerlant 4, P.O. Box 40 8200 AA LELYSTAD The Netherlands
VAN SIMAEYS, L.	Université Catholique de Louvain 3 Place des Sciences B - 1348 LOUVAIN LA NEUVE
VAN SLYPE, G.	Bureau Marcel van Dijk Avenue Louise 409, Bte 1 B - 1050 BRUXELLES
VAN VELZE, P.	Commission of the European Communities Direction-General D.G. XIII-B Bâtiment Jean Monnet Kirchberg LUXEMBOURG
VAN WIJNGAARDEN, M. (Mrs)	European Brewery Convention P.O. Box 510 - 2380 BB ZOETERWOUDE The Netherlands
VERSCHOOR, C.D.	Kluwer Rechtswetenschappen Postbus 23 7400 GA DEVENTER The Netherlands

VIAL, M. (Mme)	Editions M.D.I. B.P. 39 F-78630 ORGEVAL Cedex
VINCENT, N.	Laboratoires d'Etudes et de Recherches Synthelabo (L.E.R.S.) Avenue Paul-Vaillant-Couturier 31 F - 92220 BAGNEUX
VIRÉ, G. (Mme)	Université Libre de Bruxelles Faculté de Philosophie et Lettres Avenue F.-D. Roosevelt 50 B - 1050 BRUXELLES
VON WEISS, A.	FH Nürnberg Technisch-wissenschaftl. Verlagsausschuss des VDE Donbühlerstr. 65/II D - 8500 NUERNBERG 16
WEIL, A. (Mme)	Cabinet J. Bernard Thomas et Associés 51 Avenue Georges Mandel F - 75016 PARIS
WEISS, I.	CIBA-GEIGY AG Klybeckstr.141 CH - 4002 BASEL
WHEATON, A.	A. Wheaton & Co. Ltd. Hennock Road EXETER EX2 HRT England
WHITTINGTON, P.R.	The Chemical Society Burlington House Piccadilly LONDON W1V 0BN England
WILLIAMS, B.J.S.	NRCd The Hatfield Polytechnic Endymion Road Annexe HATFIELD Hertfordshire AL10 8AU England

WINSBURY, R.	FINTEL Limited 1 Pudding Lane LONDON EC3R 8AA England
WOLFENDEN, Lord	Chairman of the U.K Committee of Copyright Owners c/o The Publishers Association 19 Bedford Square LONDON WC1B 3HJ England
WOOD, N.	National Board for Science and Technology Shelbourne House Shelbourne Road DUBLIN 4 Ireland
ZAHORKA, H.-J.	Libertas-Europ. Zeitschrift für Christlich-Demokratische Politik Stuttgarter Str. 72 D - 7032 SINDELFINGEN 6
ZIENAU, D.	Westdeutsche Allgemeine Zeitung Gutenberg Strasse 64 D - 4600 DORTMUND 1
ZINTEL, T.	Bundesverband Druck E.V. Abteilung Bildungspolitik Postfach 1869 D - 6200 WIESBADEN 1

www.ingramcontent.com/pod-product-compliance
Lightning Source LLC
LaVergne TN
LVHW010548300926
849162LV00002B/267